FROM THE LIBRARY OF
JOHN McCARTHY
(1927 - 2011)

COMPANION TO CONCRETE MATHEMATICS

COMPANION TO CONCRETE MATHEMATICS

Mathematical Techniques and Various Applications

Z. A. MELZAK

Department of Mathematics
University of British Columbia
Vancouver, Canada

A WILEY-INTERSCIENCE PUBLICATION

JOHN WILEY & SONS   New York · London · Sydney · Toronto

Library of Congress Cataloging in Publication Data:

Melzak, Z. A.   1926–
    Companion to concrete mathematics.

    " A Wiley-Interscience publication."
    Bibliography: p.
    1. Mathematics—1961–   I. Title.

    QA37.2.M44   510   72–14171
    ISBN O–471–59338–9

Printed in the United States of America

10 9 8 7 6 5 4 3 2 1

MARGUERITAE UXORI ATQUE MARGARETAE MATRI

Of the many forms of false culture, a premature converse with abstractions is perhaps the most likely to prove fatal to the growth of a masculine vigour of intellect.

George Boole (1815–1864)

PREFACE

This book was started a long time ago as a small private collection of calculus problems, other than the usual dismal items on the volumes of solids of revolution or the rates at which ships at sea drift apart. Its scope grew upon the conviction that separating calculus from geometry, combinatorics, probability, etc., is decidely unhealthy and that it is downright sinful to teach the abstract before the concrete. It grew further on noticing the current low ebb of mathematical ingenuity and the corresponding glorification of mathematical machinery. Finally, it was staggering to discover how ignorant some brilliant mathematicians were concerning certain simple things. So, for instance, a world-famous analyst was ignorant of Euler's product for $(1 - x)^{-1}$ and a world-famous topologist could not prove that conic sections are sections of a cone. These two men were so obviously brilliant and such powerful mathematicians that they could well afford to be ignorant of lesser stuff. But, one suspects, their students may be less brilliant and powerful, and these students might pass on their ignorance and the associated contempt to their students. In justice to the two men it ought to be added that they at least were not contemptuous of what they did not know.

It was then that the final character of this book suggested itself to me: a collection of some body of ordinary but attractive mathematics which would supplement standard courses and texts by stressing concreteness, formal manipulation, intuitive appeal, and ingenuity, by using physical analogies, encouraging problem formulation, and supplying problem-solving methods. The material was then enriched by adding sketchy introductions to such mild esoterica as integral geometry, asymptotic analysis, Liouville's theory of the complexity of elementary functions, etc., and by inserting some brief historical references. I believe that the fragmentation process is so far gone in mathematics that it may be good to sacrifice much, even in rigor, for the sake of anything which unifies seemingly distant or dissimilar subjects. This might perhaps explain some odd and sudden jumps in the text: from the isoperimetric problem of the circle to measures of transcendency, or from fractional iterates to Soddy's formula for inscribed circles and Hilbert's fifth problem. For the same reason, light and loose reference is made to a few useful "principles": telescoping cancellation, minimum perturbation principle, the principle of computing the same thing in two different ways, infinite crowding principle, symbol reification principle, etc.

If there is too much reference to my own work (for the book is not all crib) it is mostly a matter of familiarity when it comes to illustrating certain tricks. Concerning this, indeed much of the book is deliberately a bag of tricks.

An author is expected to outline the level of knowledge necessary for reading his book, and to indicate for whom it is intended. Since this book deals with topics of quite different orders of difficulty, from number-multiplication games and other mathematics of the recreational type, to the zeros of the Riemann zeta function and the presumed transcendency of Euler's constant, I can only suggest that perhaps different classes of readers could use this book, and for different purposes. I hope that relative beginners and mature practitioners, teachers and students, pure mathematicians as well as the applied ones, may find something of interest in this book. As to the minimal level of preparation, this does not go beyond some real and complex analysis, and the rudiments of geometry, number theory, probability, and linear algebra. At any rate, mathematical maturity, whatever that is, though as always supremely useful, is not presupposed.

Finally, there is the pleasant duty of acknowledging many debts. First, there are my teachers. C. Fox, E. Rosenthall, and W. L. G. Williams have taught me sound elementary mathematics; W. Hurewicz and H. J. Zassenhaus have tried to teach me modern mathematics; N. Levinson has shown me effective mathematics; and S. M. Ulam has exposed me to beautiful mathematics. I have benefited greatly from my colleagues at the Bell Telephone Laboratories, especially from S. O. Darlington, E. N. Gilbert, E. F. Moore, and H. O. Pollak. It was a pleasure to be associated with my students: G. Allard, R. S. Booth, E. J. Cockayne, R. Main, J. W. Mosevich, and others; some of the material of this book has been collected or developed by them. My colleagues A. Adler, D. W. Boyd, P. C. Gilmore, J. Goldman, R. W. Hamming, J. Kennedy, J. E. Lewis, D. J. Newman, and others, have at various times offered suggestions and corrigenda. J. Spouge has spent a summer proofreading the manuscript and checking the contents. E. M. McLennan has helped with the preparation of notes for an informal problem seminar; these notes served as a basis for the manuscript. B. Kilbray and M. Swan have typed and retyped the manuscript. The staff of John Wiley & Sons have been helpful in preparing this book for print. The National Research Council of Canada has given support in the form of grants during the preparation of this book.

I thank all these. But my special thanks go to Harry Hochstadt and John Riordan for their time, interest, and generosity.

Vancouver, Canada   *Z. A. Melzak*

CONTENTS

convex body. Symmetrization.   13. *Curves in n dimensions*. Frenet–
Serret formulas and curvatures. Curves convex in *n* dimensions.

Chapter 2 ITERATION   51

1. *Preliminaries*. Definitions and simple examples.   2. *Explicit
iteration and conjugacy*. Methods for explicit iteration. Abel and
Schroder functions.   3. *Duplication and triplication*. iteration and
algebraic addition theorems. Cubic and quartic equations. Solution
of quintic equations by elliptic modular functions.   4. *Extension to
noninteger iteration*. Existence and nonuniqueness of fractional
iterates. Hilbert's fifth problem. Γ-function. Fractional derivatives.
5. *Newton–Raphson method and iteration*. Explicit expressions for
the convergents in a quadratic equation.   6. *Approximate iteration*. The *n*th iterate of the sine function.   7. *Multiple recursions*.
Iteration for systems. Arithmetic geometric mean of Gauss. A
special integral. Fast convergence of iterative methods.   8. *An
application to probability*. Probabilities in a simple fissioning process.   9. *Iteration of a function of several variables*. Partial iterates.
Soddy's formula for inscribed circles.   10. *Iteration and orders of
magnitude*. The Ackermann function. Related continuous analogue.
Primitive recursive functions.   11. *Iteration and reliable circuits*.
Moore–Shannon method of synthesis of reliable circuits our of
faulty components.

Chapter 3 SERIES AND PRODUCTS   81

1. *Telescoping cancellation*. Methods of summing certain series.
Convergence acceleration. Transformations of Kummer and Markov. Digamma function. Goldbach–Euler series. Inversion problem.
Rational approximation to the square-root function.   2. *Euler's
identity and Vieta's formula*. Euler's product for sin x/x. Vieta's
formula for π.   3. *Euler's product and Cantor's theorem*. Euler's
product for $(1 - x)^{-1}$. Cantor's theorem and irrationality.
4. *Telescoping coincidence*. Generation of periodic, doubly periodic,
and automorphic functions.   5. *Summation of certain series*. Use
of differential and integral operators. Multisection of series. Uses
of residue theorem.   6. *Products and factorization*. Evaluation of
finite products. Trigonometric factorization. Darboux method for

Chapter 4 MISCELLANEOUS ELEMENTARY TOPICS

both ends. Problems solved by working backward from the end.
Problems solved by working from both ends. Graphs and graph-
minimization. 10. *Computing* π. Brief history. Several series,
products, and continued fractions for π. Fast-convergent formulas
for π and $e^{-\pi}$. 11. *The generalized firing squad synchronization
and the labyrinth problems.* Original firing squad synchronization
problem on a line. Generalization to n dimensions. Connections
with pattern perception and computational topology. Local and
global properties. Graphs and the labyrinth traverse problems.

Mergelyan–Wesler theorem. Infinite cloud problem. Abel–Dini
theorem.   5. *Applications of certain special functions.* The skin
effect. Legendre polynomials and the zeta function. Hadamard
product and combinatorial applications. Lattice points in plane
regions. Bessel functions and Hardy's identity.   6. *Continuation
principle.* Analytic continuation, semigroups, and summability. A
general geometric continuation. Applications to communication.
Prediction problem for motion with curvature limitations. Pursuit
problems.   7. *Asymptotic analysis.* Generalities. Stirling formula.
Cesaro's theorem on divergent series and its applications. Approximate magnitude of the partition function. Polynomial recursion
and asymptotic polynomial recursion. Connections with transcendency problems. Presumed transcendency of Euler's constant.
8. *Coincidences, forbidden configurations, and hypergraphs.* Graphs
and hypergraphs. A general problem in combinatorial-geometric
probability. An application to a hard-sphere model in statistical
mechanics. Inclusion-exclusion principle. Coincidences and multiple
coincidences. Cluster integrals. Inclusion-exclusion principle for
multiindexed sets. Hypergraphs and index-sharing types.

1

GEOMETRY

1. CONIC SECTIONS

As their collective name implies, ellipses, parabolas, and hyperbolas are plane sections of circular cones. This may be observed by tilting a lit flashlight with a conical reflector at various angles against a plane wall. It can be demonstrated by some simple geometry and the demonstrations are good exercises in spatial visualization.

For the case of the ellipse we let the plane P cut the cone C as shown in Figure 1. Let S_1 be the sphere inscribed into C and tangent to P from above and let F_1 be the point of tangency. Similarly, let S_2 be the sphere inscribed into C and tangent to P from below, say at F_2. Let C_1 and C_2 be the circles of contact of the spheres with the cone. Let X be an arbitrary point on the curve E in which P cuts C. Let the straight line from the vertex of the cone through X cut C_1 and C_2 at A_1 and A_2. Since all the tangents drawn from a point to a sphere are of equal length, we have $XF_1 = XA_1$ and $XF_2 = XA_2$. Hence $XF_1 + XF_2 = A_1A_2$ which is constant. Therefore E is an ellipse with the foci F_1 and F_2.

Similar construction proves the focal property of the parabola. Let the plane P cut the cone C and let it be parallel to a generator of C. Again, let S be the sphere inscribed into C and tangent to P from above at F. Let C_1 be the circle of contact of S and C and let D be the line in which P cuts the plane of C_1. Proceeding much as before, we show that if X is any point of the curve in which P cuts C then XF is the distance from X to D. Hence the curve of intersection is a parabola with the focus F and directrix D.

For the hyperbola we use the complete double cone and we let the plane P intersect both parts. The spheres S_1 and S_2 are inscribed into the cone, one in each part, tangent to P at F_1 and F_2; the circles C_1 and C_2 are as before. If X is any point of the curve of intersection of P and C then the difference $XF_2 - XF_1$ is constant and is, in fact, equal to plus or minus the distance

1

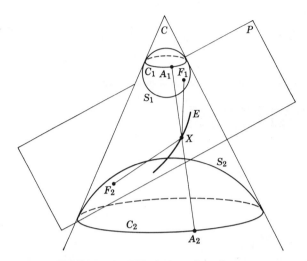

FIGURE 1.   **Elliptical section of a cone.**

between the circles C_1 and C_2 measured along any generator of C. The sign will be plus for one branch ($=$ intersection of P and one part of the cone) and minus for the other one. Hence the intersection is a hyperbola with the foci F_1 and F_2.

By tilting the plane P at various angles and moving it closer to the vertex of the cone or further away from it, the reader may convince himself that any ellipse occurs as a plane section of any given cone. This is not the case with hyperbolas: for a fixed cone only those hyperbolas whose asymptotes make a sufficiently small angle occur as plane sections.

2. AREAS ON A SPHERE

Let S be a sphere and C its enveloping cylinder touching it along the horizontal equator E, as shown in Figure 1a. If n and s are the north and the south poles of S then the line A through n and s is an axis of both S and C. For every point x on S there is a unique nearest point a on A; if x is neither n nor s then ax is horizontal and it can be produced to cut C at y. The exceptional role of n and s will present no difficulties. We call y the axial projection of x. Under this axial projection P any region R_S on S corresponds to a unique region R_C on C, and vice versa. We write $P: R_S \to R_C$ or $R_C = P(R_S)$.

Suppose that R_1 is a small rectangle $y_1 y_2 y_3 y_4$ on C, with $y_1 y_4$ and $y_2 y_3$ parallel to A, as shown in Figure 1b. If $R_1 = P(R_2)$ then R_2 is a small quadrilateral $x_1 x_2 x_3 x_4$ on the sphere, bounded by two great circle arcs $\widehat{x_1 x_4}$ and $\widehat{x_2 x_3}$, and two small circle arcs $\widehat{x_1 x_2}$ and $\widehat{x_3 x_4}$. We have Area $(R_1) =$

$\widehat{y_1 y_2} \cdot y_1 y_4$; Area $(R_2) \simeq \widehat{x_1 x_2} \cdot \widehat{x_1 x_4}$; $\widehat{x_1 x_2}/\widehat{y_1 y_2} \simeq y_1 y_4/\widehat{x_1 x_4} \simeq \cos \theta$, where θ is the latitude of x_1 on S. The approximate equalities are in the sense that the ratio of the two sides tends to 1 as the sizes of R_1 and R_2 shrink to 0. It follows that Area $(R_1) \simeq$ Area (R_2) so that the axial projection P preserves the elementary areas. Hence, by decomposition and passage to the limit, as usual in definite integrals, the axial projection preserves areas of arbitrary figures:

$$\text{Area } (R_C) = \text{Area } (R_S). \tag{1}$$

Taking C for R_C and S for R_S Archimedes proved, in essentially this way, that the area of the sphere is equal to the lateral area of the enveloping cylinder, $4\pi r^2$.

The actual proof of Archimedes [2] was somewhat different in detail. The sphere was bracketed between two surfaces of revolution obtained by rotating the regular n-gons inscribed and circumscribed to a great circle of the sphere. Certain geometrical propositions were then proved, relative to spherical and conical zones, and the final conclusion about the equality of the areas of the sphere and the cylinder was obtained by employing a special case of the squeeze principle: if $f(x) \le F(x) \le g(x)$ and $\lim_{x \to a} f(x) = \lim_{x \to a} g(x) = L$ then $\lim_{x \to a} F(x) = L$.

Archimedes must have valued this proof very highly since a cylinder enclosing a sphere was engraved on his tombstone. Some 150 years later, when Marcus Tullius Cicero was a quaestor in Sicily, the site of Archimedes' tomb was unknown and Cicero was able to find it by following the cylinder-and-sphere tradition. This is not the only example of a mathematician's work carved on his tomb. The Swiss mathematician J. Bernoulli, impressed by some geometrical properties of the logarithmic spiral, arranged to have it engraved

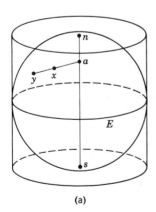

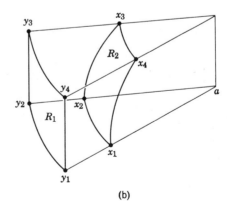

(a)   (b)

FIGURE 1.   Cylinder and sphere.

on his tombstone, together with the Latin phrase "eadem mutata resurgo"—"however changed, I return". The Dutch mathematician Ludolph van Ceulen spent many years of his life calculating π to thirty-five decimal places; this fact, together with the value of π, was attested on his tombstone in a church in Leyden.

Using the theorem (1) of Archimedes, we can find the steradian content of certain cones. The steradian content generalizes to three dimensions the radian measure of angles. To find the radian measure of an angle, as in Figure 2a, we draw a circle of radius r about the vertex o, and we divide the arc length s by the radius r. Similarly, given a cone K, as in Figure 2b, we describe about its vertex o a sphere S of radius r; the steradian content $M(K)$ is the area of the figure cut out of S by K, divided by r^2.

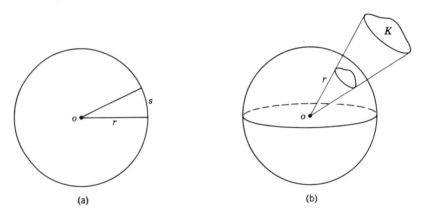

(a)   (b)

FIGURE 2.   Radians and steradians.

We consider now a rectangular cone B illustrated in Figure 3a; it is bounded by four planes passing through a point o as shown. The steradian content of this, and other cones, is of some interest in connection with antenna beams and illumination studies. We need the area of the spherical quadrilateral Q, bounded by four great circle arcs, which is cut by B out of the sphere S of radius r about o. By (1) the area of Q is equal to the area of the axial projection of Q onto any cylinder C which envelops S. We arrange C so that E, the tangent equator in which C touches S, bisects Q as in Figure 3a; we also consider the corresponding vertically bisecting meridian M which cuts E at F. The projection Q_1 of Q onto C is shown in Figure 3b and we find by some simple trigonometry that

$$\tan \gamma = \tan \beta \cos \phi. \tag{2}$$

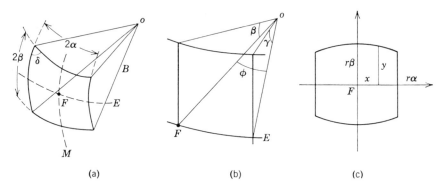

FIGURE 3. Rectangular cone.

Therefore, when Q_1 is unrolled onto a plane and F is taken as the origin, we have a region symmetric with respect to both axes, as shown in Figure 1c, and since

$$x = r\phi, \qquad y = r \sin \gamma,$$

we have by (2)

$$y = r \tan \beta \cos \frac{x}{r} \left(1 + \tan^2 \beta \cos^2 \frac{x}{r} \right)^{-1/2}.$$

Hence

$$\text{Area}(Q) = 4 \int_0^{r\alpha} y \, dx = 4r^2 \text{ arc sin } (\sin \alpha \sin \beta) \tag{3}$$

and so the steradian content of B is

$$M(B) = 4 \text{ arc sin}(\sin \alpha \sin \beta).$$

We can also determine the angle δ of the spherical quadrilateral Q shown in Figure 3a. For that purpose we show first that if T is a spherical triangle, shown in Figure 4, on a sphere S of radius r, then

$$\text{Area}(T) = r^2(\alpha + \beta + \gamma - \pi). \tag{4}$$

The quantity $\alpha + \beta + \gamma - \pi$ is called the spherical excess of T. By a lune $L(\phi)$, $0 < \phi < \pi$, we understand the smaller of the two parts in which the sphere S is divided by two meridians meeting at an angle ϕ. When the sides AB and AC are produced on S till they meet again (at the point antipodal to A) we get such a lune $L_1(\alpha)$. Let $L_2(\alpha)$ be the lune antipodal to $L_1(\alpha)$; we get similarly the lunes $L_1(\beta)$, $L_2(\beta)$, $L_1(\gamma)$, $L_2(\gamma)$. It is now not hard to show that the six lunes together cover all of S, every point of S is covered exactly once

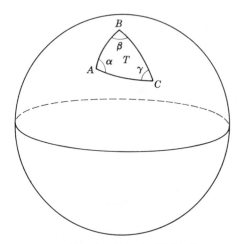

FIGURE 4.   Spherical triangle.

except for the points of T and of the triangle T_1 antipodal to T; these two triangles are covered triply. Since the area of $L(\alpha)$ is $2r^2\alpha$ we have therefore

$$2[\text{Area}(L(\alpha)) + \text{Area}(L(\beta)) + \text{Area}(L(\gamma))] = \text{Area}(S) + 4\,\text{Area}(T)$$

which gives us (4).

Next, by decomposition into spherical triangles, we find that if P is a spherical polygon of n sides and vertex angles $\alpha_1, \ldots, \alpha_n$, then

$$\text{Area}\,(P) = r^2\left[\sum_{i=1}^{n} \alpha_i - (n-2)\pi\right]. \tag{5}$$

Returning to our problem of finding the angle δ of the spherical quadrilateral Q of Figure 3a, we find by (5) that its area is

$$r^2(4\delta - 2\pi),$$

and equating this to (3) we have

$$\delta = \frac{\pi}{2} + \text{arc sin}(\sin \alpha \sin \beta).$$

We observe that the spherical-excess formula (4) is a very special case of the Gauss-Bonnet theorem from differential geometry [71] which asserts the following: let R be a simply connected region on a sufficiently smooth surface S, bounded by a finite number n of smooth arcs $C_1, C_2, \ldots, C_n$ as shown in Figure 5; let ∂R denote the oriented boundary of R, that is, $C_1 + C_2 + \cdots + C_n$, let θ_i be the exterior angles as shown, let $k_g(s)$ denote the geodesic curvature along C expressed as a function of arc-length on C, let G be the Gaussian curvature of S and dA the area element on S, then

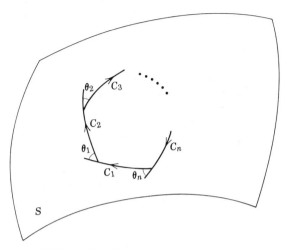

FIGURE 5.   The Gauss-Bonnet theorem.

$$\int_{\partial R} k_g \, ds + \int_R G \, dA + \sum_{i=1}^{n} \theta_i = 2\pi. \qquad (6)$$

In the special case when each arc C_i is a geodesic we have $k_g = 0$, and if further the Gaussian curvature G is constant, then

$$G \text{ Area } (R) + \sum_{i=1}^{n} \theta_i = 2\pi.$$

If $n = 3$ and S is a sphere of radius r, then $\theta_1 = \pi - \alpha, \theta_2 = \pi - \beta, \theta_3 = \pi - \gamma$, $G = r^{-2}$ and so we obtain (4).

3. LINES IN A TRIANGLE

The three medians of a triangle intersect in one point, as do the three angle bisectors and also the three heights. All three propositions follow simply from the theorem of Ceva [16]: the three segments AX, BY, CZ drawn for the triangle ABC shown in Figure 1 intersect in one point if and only if

$$AZ \cdot BX \cdot CY = ZB \cdot XC \cdot YA.$$

This in turn is a special case of a theorem of Routh [16] illustrated in Figure 2: let

$$BX : XC = l, \qquad CY : YA = m, \qquad AZ : ZB = n,$$

then

$$\frac{\text{Area } \Delta LMN}{\text{Area } \Delta ABC} = \frac{(lmn - 1)^2}{(lm + l + 1)(mn + m + 1)(nl + n + 1)}. \qquad (1)$$

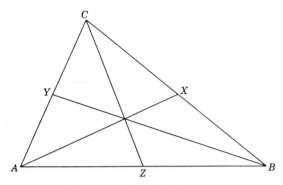

FIGURE 1.   Ceva's theorem.

This may be proved as follows. Denoting by UV the length of the straight segment and by UVW the area of the triangle, we have

$$LMN = ABC - ACX - BAY - CBZ + CNX + ALY + BMZ$$

and

$$CNX = ACX\frac{NX}{AX}, \qquad BMZ = CBZ\frac{MZ}{CZ}, \qquad ALY = BAY\frac{LY}{BY}.$$

Further

$$ACX = \frac{1}{l+1}\,ABC, \qquad CBZ = \frac{1}{n+1}\,ABC, \qquad BAY = \frac{1}{m+1}\,ABC.$$

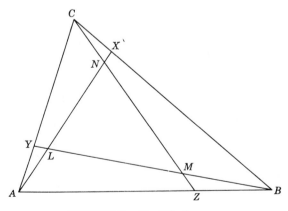

FIGURE 2.   Routh's theorem.

Therefore

$$\frac{LMN}{ABC} = 1 - \frac{1}{l+1}\left(1 - \frac{NX}{AX}\right) - \frac{1}{n+1}\left(1 - \frac{MZ}{CZ}\right) - \frac{1}{m+1}\left(1 - \frac{LY}{BY}\right). \quad (2)$$

It is now necessary to compute the three ratios

$$\frac{NX}{AX}, \quad \frac{MZ}{CZ}, \quad \frac{LY}{BY}.$$

We notice first that our whole configuration of Figure 2 may be projected orthogonally on any other plane and in the projected configuration all the corresponding ratios (l, m, n, and the three above) will be the same as in the original one. Moreover, we can project the projection itself again. Therefore, to calculate, say, NX/AX, we arrange the projections so that ABC becomes an isosceles right-angled triangle, with the right angle at B. Now some simple trigonometry shows that

$$\frac{NX}{AX} = \frac{1}{(l+1)(n+1) - l}$$

and analogously

$$\frac{MZ}{CZ} = \frac{1}{(m+1)(n+1) - n}, \quad \frac{LY}{BY} = \frac{1}{(l+1)(m+1) - m}.$$

Substituting into (2) and simplifying yields (1).

4. ANGLE MULTISECTION

It is known that the classical ruler-and-compasses constructions do not suffice to trisect an arbitrary angle, much less to divide it into an arbitrary number n of equal parts. Certain special means will accomplish this, and we describe one example. By repeated bisections and duplications we may assume that the angle θ, to be divided into n equal parts, satisfies $0 < \theta < \pi/2$. Let C be a curve with the polar equation $r = f(\theta)$ and the following property illustrated in Figure 1. For any $\theta(0 < \theta < \pi/2)$, let P be the point of C corresponding to θ, and let Q be its projection on the Y-axis. Divide OQ into n equal parts (in the figure $n = 5$) by the points $Q_1, Q_2, \ldots$, and let each Q_i be projected horizontally onto C as P_i. We wish to determine $r(\theta)$ so that for every n and θ all the angles $P_i OP_{i+1}$ are equal. This gives us the condition

$$\frac{k \sin \theta}{n} r(\theta) = r\left(\frac{k\theta}{n}\right)\sin\frac{k\theta}{n}, \qquad k = 0, 1, \ldots, n, \quad (1)$$

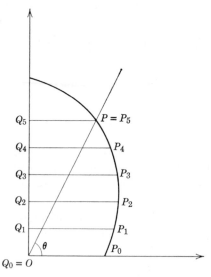

FIGURE 1.   Angle multisection.

which must hold for every integer $n \geq 1$. It is possible to *guess* from (1) that $r(\theta) = c(\theta/\sin \theta)$. It is also possible to *solve* (1) as a functional equation for $r(\theta)$. We first rewrite it as

$$r(\theta) = r\left(\frac{k\theta}{n}\right) \frac{n \sin \dfrac{k\theta}{n}}{k \sin \theta}, \qquad (2)$$

we fix k, and we use the telescoping cancellation by writing in (2) $(k/n)\theta$, $(k/n)^2\theta, \ldots, (k/n)^{m-1}\theta$ in place of θ and multiplying the resulting equations. This yields

$$r(\theta) = r\left(\left(\frac{k}{n}\right)^m \theta\right) \frac{\sin\left(\dfrac{k}{n}\right)^m \theta}{\left(\dfrac{k}{n}\right)^m \sin \theta}.$$

Passing to the limit as $m \to \infty$ and observing that $\lim \sin x/x = 1$, as $x \to 0$, we have

$$r(\theta) = r(0) \frac{\theta}{\sin \theta}. \qquad (3)$$

However, the same purpose may be accomplished somewhat differently by using the Archimedean spiral

$$r(\theta) = a\theta. \qquad (4)$$

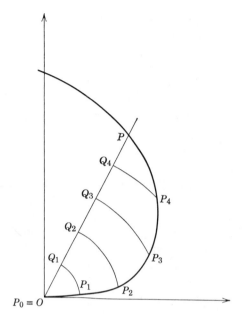

FIGURE 2.   Another multisection method.

Given an angle θ, as shown in Figure 2, we let P be the corresponding point on the curve and we divide the radius vector OP into n (in the figure $n = 5$) equal parts with the points Q_1, Q_2, Q_3, Q_4. Now we determine the points P_i on the curve by drawing circles through Q_i centered at O. It follows from (4) that the rays OP_i divide the angle $\theta = \sphericalangle XOP$ into n equal parts.

Either of the curves (3) and (4) may be used to construct a mechanical device for angle multisection.

5. PRISMOIDAL FORMULA

Let S be a (three-dimensional) polyhedral solid all of whose vertices lie in one or the other of two parallel planes P_1 and P_2. Let the base of S, in P_1, have the area A_1; let the top of S, in P_2, have the area A_2; let A_3 be the area of the section of S by the plane midway between P_1 and P_2; finally, let h be the distance between P_1 and P_2. Then the prismoidal formula states that

$$\text{Volume } S = \frac{h}{6}(A_1 + 4A_3 + A_2). \tag{1}$$

To prove this it suffices to show that (1) holds for the special case when S is a tetrahedron, for in the general case S can be decomposed into a finite union of such tetrahedra. Now, for a tetrahedron (1) is easily shown to hold.

The prismoidal formula is closely related to the idea behind the Simpson rule: let $A(x)$ be the area of the section of S by a plane parallel to P_1 at the distance x above it, then

$$\text{Volume } S = \int_0^h A(x)\, dx.$$

Since by the hypothesis $A(x)$ can be shown to be quadratic in x we have

$$\int_0^h A(x)\, dx = \frac{h}{6}\left[A(0) + 4A\left(\frac{h}{2}\right) + A(h)\right] \tag{2}$$

from which one easily deduces the Simpson rule.

By an approximation argument it may be shown that the prismoidal formula also holds for truncated cones (not necessarily circular), one-sheeted hyperboloids, both circular and elliptic, certain solids composed of parts of all these, etc. The sole requirement is that the solid S in question should be arbitrarily well approximable by a polyhedral solid satisfying the original conditions.

The equation (2) which is behind the Simpson rule has a *linear* analogue

$$\int_0^h A(x)\, dx = \frac{h}{2}[A(0) + A(h)] \tag{3}$$

valid for $A(x)$ linear in x as well as a whole series of generalizations from which one derives the Cotes rules. For instance, for $A(x)$ which is a cubic in x we have

$$\int_0^h A(x)\, dx = \frac{h}{8}\left[A(0) + 3A\left(\frac{h}{3}\right) + 3A\left(\frac{2h}{3}\right) + A(h)\right], \tag{4}$$

for $A(x)$ which is a quartic in x

$$\int_0^h A(x)\, dx = \frac{h}{90}\left[7A(0) + 32A\left(\frac{h}{4}\right) + 12A\left(\frac{h}{2}\right) + 32A\left(\frac{3h}{4}\right) + 7A(h)\right] \tag{5}$$

and so on [66]. Each of the formulas like (3), (2), (4), (5), etc., leads to a type of prismoidal formula for volumes; these are valid for larger and larger classes of solids.

6. INVERSION

Let C be a fixed circle of center O and radius r; then any point P in the plane of C, other than O, has a unique image P_1 under inversion: P_1 is that point on the ray from O through P, for which $OP \cdot OP_1 = r^2$. As is well known, straight lines and circles invert into straight lines and circles, in particular the lines and circles not passing through O invert into circles, all others invert into lines.

There are graphical aids to perform the inversion of a given locus; one of these is the Peaucellier cell, or linkage, of Figure 1a. This consists of two equal rods OA and OC and four other equal rods AB_1, AB_2, CB_1, CB_2, hinged together as shown in the figure. Since $a \sin \alpha = b \sin \beta$ we have

$$OB_1 \cdot OB_2 = a^2 \cos^2 \alpha - b^2 \cos^2 \beta = a^2 - b^2;$$

it follows that if O is fixed and B_1 traces out a locus then B_2 traces out the inverse locus. Another such inversive linkage is the Hart antiparallelogram consisting of four rods equal in pairs, hinged together as in Figure 1b. If a straight line cuts AB in O, AD in B_1, and BC in B_2, then we have the inversive property $OB_1 \cdot OB_2 = $ constant.

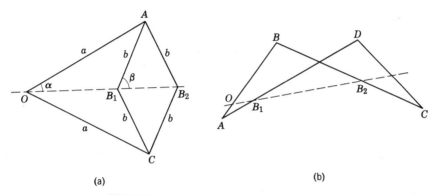

FIGURE 1. Peaucellier and Hart linkages.

Both these linkages, by fixing O and moving B_1 on a circle through O, serve as devices converting circular motion into rectilinear motion. Another way to accomplish this is based on a property of hypocycloids: if a circle C_1 rolls without slipping inside a circle C_2 of twice the radius, then any point on C_1 traces out a diameter of C_2.

However, none of these arrangements converts *uniform* circular motion into *uniform* linear motion, as may be necessary in certain mechanisms. For instance, some such device may be needed in certain sewing or spinning machines, to ensure that the thread winds onto a bobbin evenly. Uniform linear motion is easily generated by referring to the spiral of Archimedes, with the polar equation $r = k\theta$, shown in Figure 2a. Let C be a plate bounded on one side by a straight segment and on the other by the arc of the spiral $r = k\theta$, $0 \le \theta \le \pi/2$. Let two such plates be fastened flat-to-flat as shown in Figure 2b and let them rotate uniformly about O. Suppose that a straight rod PS (whose extension would pass through O) is kept in contact with the rotating plates by a spring arrangement. We suppose that PS can move back

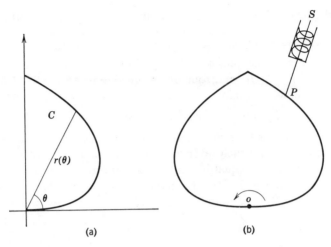

FIGURE 2.   **Archimedes spiral and uniform motion.**

and forth but not laterally. Then, as the plates rotate, P moves back and forth uniformly on a straight segment.

The inversion with respect to a fixed sphere can be defined in three dimensions just as in the plane with respect to a fixed circle. There are here similar properties concerning the inversion of spheres and planes into spheres and planes. An interesting question concerning the theory of functions of a complex variable hinges on a property of inversions in three dimensions. An analytic function $f(z) = u + iv$ of a complex variable $z = x + iy$ may be regarded as a pair $u(x, y)$, $v(x, y)$ of two real-valued functions of two real variables x and y, connected by the Cauchy-Riemann equations $u_x = v_y$, $u_y = -v_x$. We may ask now: are there any analogous analytic triples $u(x, y, z)$, $v(x, y, z)$, $w(x, y, z)$ for three dimensions? In other words, is there a theory of functions of a "three-dimensional" complex variable? The answer depends, of course, on what properties of analytic functions we wish to preserve. A most useful property here is that of conformality: if two smooth curves intersect at z_0 at an angle α then their images under an analytic function $f(z)$ intersect at $f(z_0)$, also at the angle α (provided that $f'(z_0) \neq 0$). Actually, conformality means that angles are preserved in sense as well as in size; mere preservation of size of angle is called the isogonal property and a function such as $f(z) = \bar{z}$ has the isogonal but not the conformal property. This f fails to be analytic.

In three dimensions the situation is quite different. Rigid motions, magnifications, and reflections are certainly conformal mappings here. But it can be shown, though the proof is not very simple, that the only other conformal mappings are inversions. Thus the three-dimensional space is conformally too poor to admit a theory of a three-dimensional complex variable of the same richness as the two-dimensional plane.

7. AREA AND VOLUME EQUIVALENCE

Two plane polygons P_1 and P_2 are called equivalent, or equivalent by finite decomposition, if P_1 can be decomposed into a finite number of elementary parts which after reshuffling make up P_2. We write then $P_1 \sim P_2$ and observe that if $P_1 \sim P_2$ then also $P_2 \sim P_1$. If the intersection of any two elementary parts is itself an elementary part, or a finite union of such parts, then the equivalence is transitive: if $P_1 \sim P_2$ and $P_2 \sim P_3$ then $P_1 \sim P_3$. This is shown by superimposing over P_2 the two decompositions resulting, after reshuffling, in P_1 and P_3.

With all triangles as elementary parts, the concept of equivalence stems from Hilbert and others who attempted to construct with it an elementary theory of area which avoids any appeal to limits and continuity (actually, it turns out that to obtain a purely discrete theory of polygonal area the concept of equivalence has to be changed to that of completion-equivalence: P_1 and P_2 are completion-equivalent if they become equivalent after an adjunction of two sequences of equivalent polygons, $P_1 + Q_1 + \cdots + Q_a \sim P_2 + Q_1' + \cdots + Q_a'$ where $Q_i \sim Q_i'$).

It is clear that if $P_1 \sim P_2$ then the areas are equal, $A(P_1) = A(P_2)$. We now prove the converse

$$\text{if } A(P_1) = A(P_2) \text{ then } P_1 \sim P_2. \tag{1}$$

Two lemmas are needed:

LEMMA 1. Triangles with common base and equal heights are equivalent.

LEMMA 2. Triangles with common angle and equal areas are equivalent.

In the proof of Lemma 1 the word "triangles" may be replaced by "parallelograms," as can be seen from Figure 1a where the corresponding elementary constituent triangles are indicated. Taking now two parallelograms $ABCD$

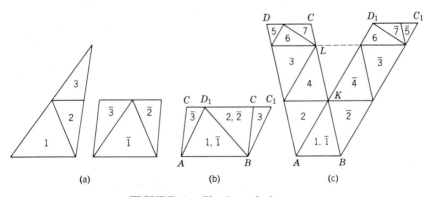

FIGURE 1. Simple equivalences.

and ABC_1D_1 we have either the case of Figure 1b, with the equivalence as indicated, or the more complicated case of Figure 1c. Here we suppose that BC and AD_1 cut in K and that $BK = KL = \cdots$ with a part LC left over. Drawing horizontal lines through $K, L, \ldots$, we now arrange the equivalence as shown.

To prove Lemma 2 we consider the two triangles ABC and AB_1C_1 as in Figure 2a. The equality of area implies $AC_1/AC = AB/AB_1$ and hence by the theorem of Thales BC_1 and B_1C are parallel. We use now Lemma 1 to show that $CBB_1 \sim CB_1C_1$ and hence also $CDC_1 \sim BDB_1$; therefore also $ABC \sim AB_1C_1$.

Now we prove our principal assertion (1). We take a polygon P of $n \geq 4$ sides and consider a vertex p. By moving it along the line through p parallel to the line joining the two neighbors of p, we move it to p_1 or p_2 as shown in Figure 2b. The number of vertices is reduced by 1; by Lemma 1 the initial and reduced polygons are equivalent. A sequence of such reductions transforms P_1 and P_2 to two triangles T_1 and T_2, with

$$P_1 \sim T_1, P_2 \sim T_2, \qquad A(T_1) = A(T_2).$$

We convert T_i to an equivalent rectangle and this to an equivalent right-angled triangle R_i. Now $R_1 \sim R_2$ by Lemma 2; this proves (1).

Similar definition of equivalence can be made in three dimensions for two polyhedra P_1 and P_2, with all polyhedra admissible as elementary constituent parts. An obvious necessary condition for equivalence is the equality of volumes: $V(P_1) = V(P_2)$. Now, a natural extension of our Lemma 1 leads to the following:

> (HQ) are polyhedral pyramids with common base and equal heights equivalent?

This was considered so important by Hilbert that he included it (in a somewhat different form) in his famous list of 23 unsolved problems, presented at

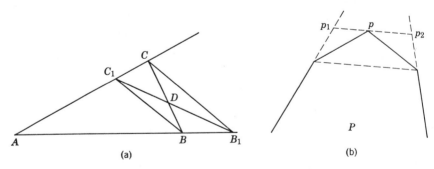

(a)   (b)

FIGURE 2.   Equivalence of triangles.

the Second World Congress of Mathematicians in Paris, in 1900. A negative answer was given within a few months by M. Dehn [18] who proved a theorem giving another, independent, necessary condition for equivalence. Dehn was motivated by the following analogy from the plane case of equivalence of two polygons. Suppose that two polygons P_1 and P_2 are equivalent, let $\alpha_1, \ldots, \alpha_n$ be the angles of P_1 and $\beta_1, \ldots, \beta_m$ those of P_2. Let $\gamma_1, \ldots, \gamma_s$ be all the angles of the constituent triangles. We apply the principle of counting one thing in two different ways, and add up γ_i's when the triangles make up P_1 and when they make up P_2. In each case we group together angles with common vertex, and we obtain

$$\sum \alpha_i + K_1 \pi = \sum \gamma_i = \sum \beta_i + K_2 \pi \tag{2}$$

since the angles at a vertex of P add up to the vertex-angle, and the angles at any other point add up to a multiple of π; K_1 and K_2 are integers. Hence

$$\sum_1^n \alpha_i - \sum_1^m \beta_i = K \pi. \tag{3}$$

For ordinary Euclidean plane this is a triviality because we know anyway that the angles of an n-gon add up to $(n - 2)\pi$. But already for spherical polygons the equivalent of (3) yields nontrivial information.

Dehn applies similar counting procedure to two equivalent polyhedra P_1 and P_2, counting the dihedral, or edge, angles. Let $\alpha_1, \ldots, \alpha_n$ be the dihedral angles of P_1, $\beta_1, \ldots, \beta_m$ those of P_2, $\gamma_1, \ldots, \gamma_s$ those of all the constituent elementary polyhedra. We again add up the dihedral angles γ_i in two ways, grouping together the dihedrals based on the same edge. However, now we do not obtain anything as simple as (2), on account of trouble with multiplicities. Instead, we get

$$N_1 \pi + \sum_1^n p_i \alpha_i = \sum_1^s q_i \gamma_i$$
$$N_2 \pi + \sum_1^m r_i \beta_i = \sum_1^s t_i \gamma_i \tag{4}$$

where $N_1, N_2, p_i, q_i, r_i, t_i$ are integers. By an involved process Dehn shows that the decompositions of P_1 and P_2 can be varied so as to ensure that the right-hand sides of (4) are equal; this yields

$$\sum_1^n p_i \alpha_i - \sum_1^m r_i \beta_i = N \pi, \tag{5}$$

which is the Dehn necessary condition for equivalence.

Consider now the two pyramids of Figure 3. Both are based on the unit square and are of height $1/2$, that of Figure 3a has its vertex over the middle of the base, that of Figure 3b has its vertex over the middle of a side of the

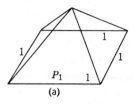 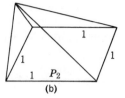

FIGURE 3. Inequivalent pyramids.

base. An explicit computation shows that (5) cannot hold for P_1 and P_2—hence a negative answer to Hilbert's question. With some elements of set theory it is possible to show further that there is a continuum of such pyramids which are pairwise inequivalent. A sufficient and necessary condition for equivalence has been given by Hadwiger; however, it is entirely nonconstructive and uses a technique similar to that of a Hamel basis.

The reader may wish to investigate the following. Call a decomposition of a polyhedron P into constituent parts $Q_1, \ldots, Q_s$ a simple decomposition if any two edges of Q_i's which have more than one point in common necessarily coincide. It is easy to show that any decomposition of P can be refined to a simple decomposition. However, is it true that if $P_1 \sim P_2$ then there exists a decomposition which is simultaneously simple over P_1 and over P_2? If the answer is yes, then the simple device of counting the dihedral angles in two ways proves directly Dehn's condition (5).

The following may also be verified. There is no difficulty with multiplicities if we perform the double counting process not on the dihedral angles γ_i, to get $\sum \gamma_i$, but if instead we form in two ways the weighted sum $\sum l_i \gamma_i$, with each dihedral angle weighted by the corresponding edge-length l_i. It is then an easy consequence to obtain the Dehn condition (5) if all edge-lengths are rational numbers. This suggests further possible connection between the equivalence problem for polyhedra, and the formula for the volume of the outer parallel body K_r to a convex body K, in three dimensions: K_r is the union of all balls of radius r centered in K. If $V(K)$, $A(K)$, $M(K)$ denote respectively the volume, area, and integral of mean curvature, of K then [6]:

$$V(K_r) = V(K) + rA(K) + r^2 M(K) + 4\pi r^3/3.$$

In the special case when K is a polyhedron with the edge-lengths $l_1, \ldots, l_s$ and the corresponding dihedral angles $\alpha_1, \ldots, \alpha_s$, the above formula reduces to

$$V(K_r) = V(K) + rA(K) + r^2 \sum_1^s l_i \alpha_i + 4\pi r^3/3.$$

We now consider, very briefly, a quite different problem which is also concerned with a decomposition of figures into parts of prescribed type.

Given a square or a rectangle, we ask whether it can be decomposed into a finite number of squares $S_1, \ldots, S_n$; such a decomposition is called perfect if no two squares S_i are of the same size, and it is called simple if no proper subset of two or more constituent squares forms a rectangle. It is clear that if the decomposition is to be simple or perfect the problem is not quite trivial, otherwise it might be. Various questions may be put now concerning minimum values of n for which simple, perfect, simple and perfect, simple though imperfect, etc., decompositions exist. Contrary to what was believed, there exist perfect simple decompositions of a square (the lowest known value of n is 38) [9] [49]. An example of a perfect simple decomposition of a rectangle into nine squares is given in Figure 4a; a simple though imperfect decomposition of a square into 13 squares is given in Figure 4b; in both figures the numbers refer to the lengths of the sides of constituent squares. Figure 4c is the graph of vertical adjacency for the 13 component squares of Figure 4b. We start from the upper edge e_1 in Figure 4b which is represented as the topmost node E_1 in Figure 4c. Two squares S_1 and S_2 adjoin e_1 and they are represented as the descending edges from E_1. S_6 and S_7 are the two squares below S_1 in Figure 4b and they are represented by the two edges descending from the node E_2 in Figure 4c. Similarly, S_3, S_4, and S_5 are below S_2 and so there are three edges descending from the node E_3. One of these, S_3, goes to the node E_2 since S_6 is below S_3. We continue in this way till we get to the node E_0 representing the lower edge e_0 of the square in Figure 4b.

Suppose now that e_1 and e_0 are metal electrodes with e_1 at the potential 1 and e_0 grounded, and that constant uniform current is flowing from e_1 to e_0. All points on a horizontal line are at the same potential; also, we may make a cut along any vertical segment without changing anything. It follows that each square S_i represents one ohm resistance because the voltage ($=$ vertical dimension of S_i) equals the current ($=$ horizontal dimension of S_i), and the

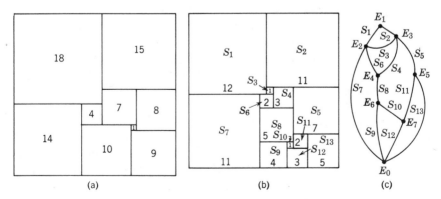

FIGURE 4.   Squarings of rectangles and squares.

13 unit resistances are connected as shown in the vertical adjacency graph of Figure 4c. Finally, the resistance of the whole network of Figure 4c is also one ohm. Therefore the decomposition of a square into squares corresponds, by way of the vertical adjacency graph, to a network of one ohm resistances, whose total resistance is also one ohm. The same holds for similar decompositions of rectangles into squares and of rectangles into rectangles, but then the various resistances will be the ratios of heights to base-lengths, rather than 1. We observe that any simple decomposition leads to a network which has no series-parallel parts; also, it is quite easy to determine any current in the network and the potential of any node.

8. ISOPERIMETRIC PROBLEMS FOR CONVEX HULLS

In this section we solve partially two problems of the isoperimetric type: to maximize an integral $I_1(f)$ depending on the unknown function f and its derivatives, while another such integral $I_2(f)$ is kept constant. We observe first two useful related principles:

(a) It is advantageous to apply those perturbations to f which keep one of the integrals constant and change the other one in the right direction, i.e., those keeping $I_2(f)$ constant and increasing $I_1(f)$ or those keeping $I_1(f)$ constant and decreasing $I_2(f)$.

(b) There is a duality of the following type: the same function which *maximizes* $I_1(f)$ while keeping $I_2(f)$ constant, also *minimizes* $I_2(f)$ while keeping $I_1(f)$ constant (see [15]).

Similar duality appears in some simple extremal problems of calculus, for instance, in the problem of determining the length L of the longest straight rod which can be carried horizontally round a right-angle bend at which two corridors of widths a and b meet. With reference to Figure 1 we find that the *longest* such L is the *shortest* length xy of the figure. (On the other hand, we may also ask about the plane region of largest area, or largest diameter, which can similarly be carried horizontally round the corner in Figure 1; these, and other similar problems concerning various maximal figures which can go through doors and around corners, are considerably harder than the simple problem illustrated in Figure 1.)

Our first problem is to find a plane closed rectifiable curve C of fixed length L which encloses the largest possible area A. It is sometimes called the problem of Dido, or the problem of Hengist and Horsa, these semimythical personages having been concerned in shady land-grant deals involving as much land as can be enclosed by a bull's hide cut into thin strips and tied together. The official mathematical name of our problem is the isoperimetric problem of the circle.

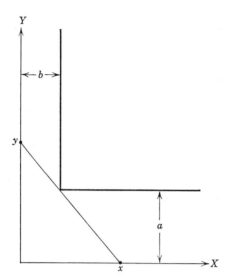

FIGURE 1. Right-angle bend.

First, we may assume that C is convex, for otherwise a perturbation may be applied which keeps L constant and increases A. This perturbation is illustrated in Figure 2 and consists of reflecting a concavity outward by mirroring it in a tangent line.

Next, let P and Q be any two points on our convex curve C which divide the circumference into two parts of equal length. Then it may be assumed that the area of C is also divided by the line PQ into two equal parts. For otherwise one half of the curve together with its reflection in PQ will enclose a larger area

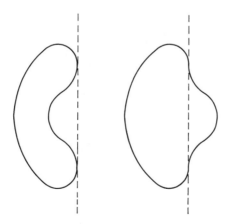

FIGURE 2. An isoperimetric perturbation.

with the same overall length. Consider now any such half C_1 of C, bounded by the points P and Q, as shown in Figure 3. Let X be any point on C_1 and let it be joined to P and Q be straight segments. To prepare for the right perturbation consider the configuration mechanically: P and Q are free to slide on their line S; X is a hinge; and in the process of sliding the arcs PX and XQ of C_1 do not change shape. Of the three component parts A_1, A_2, A_3 of the area enclosed by C_1 against S, only A_2 changes in the process, while the length of C_1 is unchanged. Hence the area enclosed is maximum when the angle α is $\pi/2$. Since X is arbitrary, every angle on PQ is a right angle and so C_1 is a semicircle and C itself a circle.

The less experienced reader may wonder about the word " partially " in the first sentence of this section. What is " partial " in the above proof? In spite of its informality it is rigorous but the " partialness " comes from the tacit assumption that there indeed exists a rectifiable plane closed curve C of length L which maximizes the area enclosed. Occasional disastrous consequences of assuming that a solution to an extremal problem exists may be observed in the following classical example: of all positive integers, 1 is greatest, for every other one is increased by squaring. In a rigorous treatment it would be necessary to supply an existence proof for such a curve C. This can be done fairly simply by an appeal to Blaschke's selection principle (see the section on convexity). As an aside, we may observe here similar situations when a mathematician is criticized by certain physicists (or chemists or engineers or economists) for being overly concerned with the problem of establishing the existence of solutions of, say, some differential equations which deal with a natural phenomenon X. The physicists say: X is real and a real X must have real behavior, so real solutions really exist. To which an answer may be given: yes, X is real but the equations describing it are almost always approximations, and thus these equations describe not X but some hypothetical simplified X_1; therefore the need for existence-proofs follows from modesty of admitting one's partial ignorance.

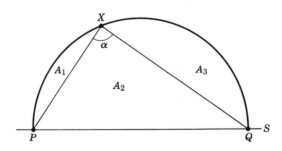

FIGURE 3.   Area perturbation.

Returning to the isoperimetric problem of the circle, we observe that what has been proved is this: *if* there exists a maximizing curve then it must be a circle. We have then, provisionally, the isoperimetric inequality: let C be a rectifiable plane closed curve, $L(C)$ its length, $A(C)$ the area enclosed by it, then

$$L^2(C) \geq 4\pi A(C) \tag{1}$$

with the equality holding if and only if C is a circle. The above suggests the possibility of an improved inequality of the type

$$L^2(C) - 4\pi A(C) \geq Q(C) \tag{2}$$

where $Q(C)$ is some quantity depending on the geometry of C, which is always positive, except for the circle when it vanishes. Briefly, we want a more precise estimate of the isoperimetric deficit. Similar problems of finding more informative measures of deficiency arise in other branches of mathematics as well. For instance, from number theory we know that the number π is transcendental; i.e., if $P(x)$ is a polynomial with integer coefficients then $P(\pi) \neq 0$. Hence arises the question of transcendency measure for π: if

$$P(x) = \sum_{j=0}^{N} a_j x^j, \qquad |a_j| \leq h$$

to find a possibly large function $M(N, h)$ which is always positive and for which

$$|P(\pi)| \geq M(N, h).$$

The existence of such $M(N, h)$ is plainly equivalent to the transcendency of π. For examples of transcendency measures, see [41].

Returning to (2), we let C be convex, $R(C)$ and $r(C)$ be the radii of the smallest circle containing C and largest circle contained in C. Then we have the Bonnesen inequality of the type (2):

$$L^2(C) - 4\pi A(C) \geq \pi^2 [R(C) - r(C)]^2.$$

We now generalize the isoperimetric problem of the circle. As with some other cases, there are generalizations which are obvious and others which are not, and the obvious-unobvious dichotomy does not necessarily coincide with the important-unimportant or the uninteresting-interesting. At any rate, the obvious, and extremely important, generalization is: to find a closed surface of fixed surface-area which encloses the largest possible volume.

We shall consider an unobvious generalization. First, as usual before a generalization, our isoperimetric problem of the circle will be restated. Since the maximizing curve C was shown to be convex, the plane region enclosed by C together with C itself forms the convex hull of C, i.e., the smallest convex region containing C. Next, we recall the simultaneous length-and-area bisection property. Now we rephrase our isoperimetric problem of the circle thus: to find a plane rectifiable arc C_1 of fixed length ($=L/2$) whose convex hull $\bar{C}_1$ has largest possible area. The answer is, as was shown, a semicircle.

By this time our intended generalization might be guessed: to find a rectifiable space arc C_1 of fixed length, which maximizes the volume of the convex hull $\bar{C}_1$.

This convex hull $\bar{C}_1$ is again defined as the smallest convex region containing C_1. Alternatively and equivalently, it is the intersection of all half-spaces containing C_1. Here a half-space is defined to be the set of all points which lie on, or to one side of, a plane. To help him with the visualization of $\bar{C}_1$ the reader may consider the following analogy: let C_1 be made of very thin, very hard, very rigid wire of very high melting point; let C_1 be dipped entirely into molten metal and allowed to solidify; when the whole mass is cool we use the *flat part* of a large grinding wheel which grinds away all spare metal but is too soft to touch the imbedded wire C_1; what remains is precisely the convex hull $\bar{C}_1$.

This time the result we obtain is very partial indeed. We assume not only the existence of the maximizing arc C_1 but also the following very special property: a plane through the end-points P and Q of C_1 and through any point X of C_1 cuts the solid convex hull $\bar{C}_1$ in the triangle PQX.

We locate C_1 with respect to a Cartesian coordinate system so that P is at the origin and Q is on the positive z-axis. Our special assumption about C_1 implies that the projection of C_1 onto the XY-plane is a simple closed curve C. Further, the same assumption implies that the volume of $\bar{C}_1$ is equal to the volume of the cone K with vertex Q and base C. Since the volume of K depends *only* on the height $h = PQ$ and on the area enclosed by C, but not on C itself, by reference to the isoperimetric problem of the circle we find that C is a circle; let its radius be r.

We now use the duality: C_1 is the *shortest* arc which has its convex hull of *fixed* volume. Since C is a circle, C_1 lies on a circular cylinder. Unrolling the latter, after slitting it along PQ, onto a plane we find that when unrolled, C_1 is a straight segment. Hence, finally, C_1 is one turn of a circular helix, its length is

$$L = \sqrt{h^2 + 4\pi^2 r^2}$$

and the volume of $\bar{C}_1$, or of K, is $(\pi/3)r^2 h$. We are thus left with a simple exercise in calculus: maximize $(\pi/3)r^2 h$ while $h^2 + 4\pi^2 r^2 = L^2$, and we find

$$\text{Vol}(\bar{C}_1) = L^3/(18\pi\sqrt{3}).$$

The (tentative) isoperimetric inequality is accordingly:

$$\frac{\text{Vol}(\bar{C}_1)}{L^3(C_1)} \leq \frac{1}{18\pi\sqrt{3}}$$

with the equality holding for C_1 which is one turn of a circular helix of pitch $1/\sqrt{2}$, and only then.

The reader may wish to prove the equality of the volumes of $\bar{C}_1$ and K. He may also wonder whether the special property of C_1 is really necessary. There is some reason to believe that it is not, and the reader may try to relax that property. Whether our solution holds under the modest assumption of mere existence of the maximizing arc C_1, appears to be still unknown.

We mention briefly the *isodiametric* problem of the circle: of all plane sets, no two points of which are further than one unit apart, to find the set X whose area $A(X)$ is largest. We suppose that such a maximizing set X exists and, as in the isoperimetric problem, we show easily that X is convex. Let L_1 be a line supporting X at O and let L be the line through O, perpendicular to L, consider polar coordinates as in Figure 4. Then we have

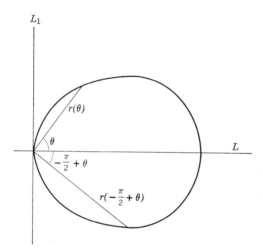

FIGURE 4. Isodiametry.

$$r^2(\theta) + r^2\left(-\frac{\pi}{2} + \theta\right) \leq 1$$

and therefore

$$A(X) = \frac{1}{2}\int_{-\pi/2}^{\pi/2} r^2(\theta)\, d\theta = \frac{1}{2}\int_0^{\pi/2}\left[r^2(\theta) + r^2\left(-\frac{\pi}{2} + \theta\right)\right] d\theta \leq \frac{\pi}{4}.$$

Hence $A(X)$ attains its maximum $\pi/4$ when X is the circular disk of unit diameter.

We can formulate a general isodiametric problem as follows. Let X be a set in the Euclidean n-space and $D(X)$ its diameter. Let $F(X)$ be a nonnegative number with the following properties:

(a) $F(X) = F(X_1)$ if X is congruent to X_1,
(b) $F(\lambda X) = \lambda^a F(X)$,
(c) $F(X) \leq F(Y)$ if $X \subseteq Y$,
(d) $F(X)$ is continuous.

Here λX is the scaled-up replica of X in the ratio $\lambda : 1$ (and $\lambda > 0$), a is a positive constant, and in (d) the continuity is taken with respect to the Hausdorff set-distance $d(X, Y)$: if X_ε is the union of all balls of radius ε centered in X, then

$$d(X, Y) = \inf\{\varepsilon : X \subset Y_\varepsilon \text{ and } Y \subset X_\varepsilon\}.$$

Examples of such functionals $F(X)$ are: the volume of X, its surface-area, and its Borsuk gauge $G(X)$. This is defined as the infimum of all numbers u such that X is a union of $n + 1$ sets of diameter $\leq u$.

Now the general isodiametric problem is: maximize $F(X)$ subject to the condition $D(X) \leq 1$. As in the Section 7 of Chapter 4, it can be shown from the monotonicity condition (c) that maximum of $F(X)$ is attained when X is adjunction-complete (any adjunction increases its diameter), i.e., of constant width.

9. REFLECTIONS AND IMAGES

A geometrical problem may be sometimes crucially simplified by a mirror reflection of certain configurations in certain lines or planes a finite or infinite number of times. A very simple example is illustrated in Figure 1: given two points P and Q, on the same side of a straight line L, to find a point T on L which minimizes the sum $PT + TQ$. Reflecting Q in L into Q_1 we find that $PT + TQ = PT + TQ_1$. Hence the minimum occurs when P, T, Q_1 are collinear and so $\alpha = \beta$. A similar though less obvious use of reflection occurs in the elementary solution of Fagnano's problem: into a given acute-angled

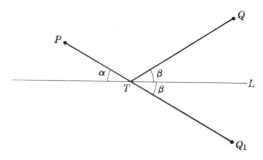

FIGURE 1.   A simple reflection.

triangle ABC to inscribe a triangle XYZ of minimum perimeter (see Figure 2.) We apply first the principle of minimum perturbation and we fix X and Y while Z varies on AB. XZ and YZ being variable, the sum $XZ + YZ$ is to be minimized; hence by the previous problem we find $\angle XZA = \angle YZB$. Similarly, by varying Y alone and then X alone we find that $\angle CYX = \angle ZYB$ and $\angle AXZ = \angle CXY$.

Now reflect Z in AC into Z_1 and in BC into Z_2. By the laws of reflection $Z_1X = ZX$ and $Z_2Y = ZY$; by the foregoing angle-equalities the points Z_1, X, Y, Z_2 are collinear. Hence the perimeter of XYZ is Z_1Z_2. Further, $Z_1C = ZC = Z_2C$ and $\angle Z_1CZ_2 = 2\angle ACB$. Hence the perimeter to be minimized is $2CZ \sin \angle ACB$. Here CZ is the only variable—it follows that Z is the projection of C into AB so that CZ is one of the heights of ABC. Similarly, AY and BX are the other two heights.

Let S be the circumference of the minimal triangle XYZ, let T be the area of the triangle ABC and R the radius of its circumscribed circle. Using the sine law

$$\frac{CB}{\sin \angle CAB} = \frac{BA}{\sin \angle ACB} = \frac{AC}{\sin \angle ABC} = R$$

we find that $S = 4T/R$.

Marquess Giulio de Toschi di Fagnano (1682–1766) and his son Marquess Gianfrancesco di Fagnano (1715–1797) were Italian mathematicians, active in geometry, analysis, and other branches of mathematics. The problem we have discussed is taken from a paper of Fagnano, Jr. (1779) in which several geometrical questions are put and solved both by calculus and by pure geometry:

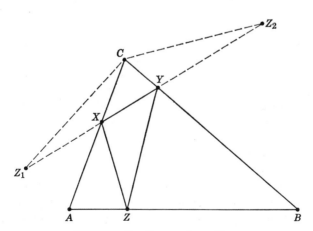

FIGURE 2. Fagnano's problem.

(a) to find a point which minimizes the sum of distances to the three vertices of a triangle (the case $n = 3$ of the Cavalieri–Steiner problem),
(b) the same for the squares of distances,
(c) the same for a quadrilateral,
(d) to inscribe a triangle of minimum circumference into a given acute-angled triangle,
(e) given an ellipse with center o and foci f_1, f_2, to find the point x of the ellipse so as to maximize the difference $\sphericalangle oxf_1 - \sphericalangle oxf_2$.

Fagnano, Sr., may be regarded as the forerunner of the theory of elliptic functions, on account of the following theorem proved by him in 1716. Let E be the ellipse with parametric equations $x = a \sin t$, $y = b \cos t$; let $X(t)$ be the point corresponding to the angle t; let $X_1 X_2$ denote the (shorter) arc of E with given end-points. We put

$$A = X(\pi/2), \qquad B = X(0), \qquad P = X(\phi), \qquad Q = X(\psi)$$

where $\tan \phi \tan \psi = a/b$. Then Fagnano proved a general theorem which implies in particular that

$$BP - AQ = (a^2 - b^2)^{1/2} \sin \phi \sin \psi;$$

this turned out later to be a special case of the addition theorem for the Jacobian elliptic function sn. When $P = Q$ the point P is called Fagnano's point F of E, its coordinates are $(a^{3/2}(a + b)^{-1/2}, b^{3/2}(a + b)^{-1/2})$, and $BF - AF = a - b$. F is also the point in which the circle inscribed into the space between E and its circumscribing rectangle (with sides parallel to the axes) touches E. Two years later, in 1718, Fagnano started his work on the lemniscate, showing that the quadrantal arc of the lemniscate can be algebraically divided into 2, 3, or 5 equal parts; this is the first example of the so-called complex multiplication for elliptic functions.

In another problem of the same type as Fagnano's we have a triangle ABC and two points p and q inside it, as shown in Figure 3. It is required to find a ray of light which is projected from p, reflects successively from AC, AB, and then BC, and then passes through q. On account of the laws of reflection the solution is obtained by reflecting ABC in AC so that p reflects into p_1; we also reflect ABC in AB, giving us the triangle ABC_1, which is then reflected in $C_1 B$ to yield $C_1 A_1 B$. Let q_1 be the reflection of q in AB, and q_2 the reflection of q_1 in $C_1 B$. We now join p_1 and q_2 by a straight segment and reconstruct the path $prstq$ of the reflected ray.

There are at most six different types of reflections possible in which every side is hit just once. By considering them the reader may wish to solve the following problem: join p to q within ABC by the shortest path which touches each side.

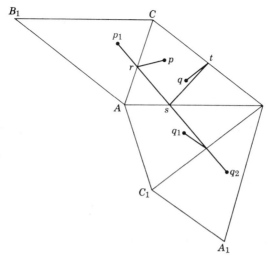

FIGURE 3.   Reflections inside a triangle.

An essential use is made of the reflection principle in the following proof of Bertrand's lemma: if in a ballot candidates P and Q obtain p and q votes, respectively, with $p > q$, then the probability that P led Q throughout the voting is $(p - q)/(p + q)$.

A convenient graphical representation of the whole voting history is the polygonal path with vertices $(0, s_0)$, $(1, s_1)$, ..., (x, s_x) where $x = p + q$, $y = s_x = p - q$, $s_0 = 0$, and $s_i = s_{i-1} \pm 1$ depending on whether the ith ballot is cast for P or for Q. An example with $p + q = 6$, $p - q = 2$ is given in Figure 4a. If N_{xy} is the total number of such ballot paths from $(0, 0)$ to (x, y) then

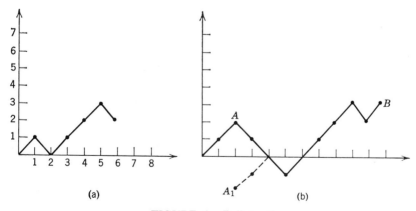

FIGURE 4.   Ballot paths.

$$N_{xy} = \binom{p + q}{q} = \binom{x}{\dfrac{x - y}{2}}$$

since there are q choices to be made for the Q-votes out of the total of $p + q$. Suppose now that A and B are two vertices of a ballot path as shown in Figure 4b, with A above the x-axis; let A_1 be the reflection of A in the x-axis and together with A let us also reflect the whole part of the ballot path between A and the earliest vertex on the x-axis, as shown in the figure. This reflection provides a 1 : 1 correspondence between all those ballot paths from A to B which have a vertex on the x-axis and all ballot paths from A_1 to B. Hence the corresponding numbers of paths are equal.

Let N be the number of paths consistent with the requirement that P leads Q throughout. N is clearly the number of ballot paths from $(1, 1)$ to (x, y) which have no vertex on the x-axis. By the foregoing, N is therefore $N_{x-1\,y-1} - N_{x-1\,y+1}$ which gives for the desired probability

$$\frac{N}{N_{xy}} = \frac{p - q}{p + q}.$$

Related use of reflections occurs in the solution of certain problems in electrostatics by Kelvin's method of images. In the simpler cases the object is to calculate the potential due to a point charge e in the presence of metal surfaces kept at 0 potential. The correct boundary condition, namely the 0 potential, may be sometimes synthesized by placing a finite or infinite number of charges $\pm e$ at the images, or repeated images, of the original point charge in the loci of the 0 potential. In the simple case of a charge e at $(a, 0)$ and the y-axis kept at 0 potential, we find that the resulting potential is the same as with e at $(a, 0)$, $-e$ at $(-a, 0)$ and no boundary conditions. Similarly, if a charge e is placed at (a, b) in the first quadrant, and the positive x-axis and y-axis are kept at 0 potential the result is the same as with four charges: e at (a, b) and at $(-a, -b)$, $-e$ at $(a, -b)$ and at $(-a, b)$, without any boundary conditions.

For a case requiring an infinity of images we consider a charge e at $(0, h)$, $0 < h < 1$, between the parallel lines $y = 0$ and $y = 1$ kept at 0 potential. By reflecting the strip between these two lines infinitely in both directions we find that the potential $V(x, y)$ at (x, y), with $-\infty < x < \infty$ and $0 < y < 1$, is that due to an infinity of charges e at $(0, 2n + h)$, $n = \ldots, -1, 0, 1, \ldots$, and an infinity of charges $-e$ at $(0, 2n - h)$, $n = \ldots, 1, 0, 1, \ldots$, without any boundary conditions. Since the potential at (x, y) due to a charge e at (x_1, y_1) is

$$\frac{e}{\sqrt{(x - x_1)^2 + (y - y_1)^2}}$$

we find that

$$V(x, y) = e \sum_{n=-\infty}^{\infty} \left[\frac{1}{\sqrt{x^2 + (y - 2n - h)^2}} - \frac{1}{\sqrt{x^2 + (y - 2n + h)^2}} \right].$$

We could similarly handle the case of a charge e placed at (x, y), with $0 < x < a$ and $0 < y < b$, inside the rectangle with vertices $(0, 0)$, $(a, 0)$, $(0, b)$, (a, b), assuming that the sides of the rectangle are kept at 0 potential. Here it is also possible to compute the resultant potential by placing charges at an infinite lattice of points which are the repeated images of (x, y). But we may also solve the problem by using a conformal mapping in which an analytic function f maps the rectangle conformally onto the upper half-plane. It will turn out that f is a Jacobian elliptic function; using the principle of calculating the same thing in two different ways, the reader may wish to equate the results and thus obtain a series expansion for the elliptic function in question. The reader may also wish to observe some connections between the above use of reflections and what has elsewhere been called the principle of telescoping coincidence. Finally, the reader may wish to consider inversion in circles and extend the method of images for charges in the presence of circular boundaries kept at 0 potential [44].

Reflection technique of the same type will enable us to solve the problem of the path of a ray of light traveling inside a square S with internally reflecting sides. Let S have the vertices $(0, 0)$, $(1, 0)$, $(0, 1)$, $(1, 1)$ and let the ray pass through the point (a, b) inside S with the direction cosines h, k; it may be assumed that $hk \neq 0$. The path of the ray is shown in Figure 5a, reflecting S in its sides and repeating the same for each successive reflection, we obtain an infinite configuration shown in Figure 5b. If by any chance the ray hits a

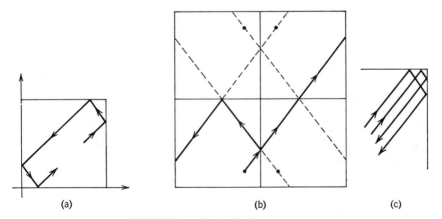

(a)   (b)   (c)

FIGURE 5.   Reflections in a square.

corner of the square, continuity argument illustrated in Figure 5c suggests that the ray be required to turn back on its path.

Consider an arbitrary point $P(x, y)$ in S; then there are four infinite sequences of the multiple images of P under reflections

$$(x + 2n, y + 2m), (2 - x + 2n, y + 2m), (x + 2n, 2 - y + 2m),$$
$$(2 - x + 2n, 2 - y + 2m); \quad (1)$$

here m and n range independently over all integers. The main point of using the reflections is that the images of the segments constituting the path form a straight line L. It is now a simple matter to show that if h/k is rational then the path is closed and periodic.

If h/k is irrational the path is not closed and it is dense in S: it passes arbitrarily closely to every point (x, y) in S. To prove this we have to show that L passes arbitrarily closely to some image of one or another of the four sequences (1) of the images of (x, y). This may be proved by an appeal to Kronecker's theorem; let the real numbers $c_1, c_2, \ldots, c_s$, 1 be linearly independent over the integers, i.e., suppose that if integers $n_1, n_2, \ldots, n_{s+1}$ satisfy

$$\sum_1^s n_i c_i + n_{s+1} = 0$$

then all the integers are 0; let (nx) denote the fractional part of nx: $(nx) = nx - [nx]$; then the Kronecker theorem asserts that the sequence

$$((nc_1), (nc_2), \ldots, (nc_s)), \qquad n = 1, 2, \ldots \quad (2)$$

is dense in the s-dimensional unit cube.

As a matter of fact, the foregoing is not the whole truth about the motion of the reflected ray, just as the Kronecker theorem is not the whole truth about the sequence (2). The Bohl–Sierpinski–Weyl theorem asserts that under the stated conditions on $c_1, \ldots, c_s$, 1 the points (2) are not only dense in the unit cube C but, further, they are equidistributed in that cube [56]. That is, if V is any sufficiently regular subset of C and $N(n)/n$ is the fraction of the first n points of (2) falling in V, then the limit $\lim N(n)/n$, as $n \to \infty$, exists and is the volume of V.

A consequence of the above is that the path of the ray in S is ergodic in S: it spends, to put it roughly, the same time in the neighborhood of any point of S. More precisely, we can express the ergodic property as follows: let the ray move with uniform speed and let V be any sufficiently regular subset of S, then the fraction of the time interval $(0, T)$ spent by the ray in V tends, as T grows, to the area of V. Thus the ergodic property appears in the form of asymptotic equality of two averages: space average (since the probability that a random point in S falls in V is the area of V) and time average. This is the form of ergodicity as it appears in physics, where "particle average" is equated to "ensemble average."

The reader may wish to attempt to prove the harder part of our proposition (h/k irrational) by contradiction: assuming that the path does not approach some point (x, y) in S to within distance ε, we conclude that L misses all the circles of radius ε about the centers (1); now we might perhaps use the techniques similar to those outlined in the section on infinite crowding of the chapter on intermediate miscellaneous topics. If successful, such proof, after generalization to n dimensions, would constitute an independent proof of Kronecker's theorem.

The reader may also wish to attempt a proof of the following conjecture: let the unit square S be placed vertically and let a point mass (a small perfectly elastic ball) be projected sufficiently fast from some point in S in a suitable (arbitrary?) direction, supposing a down-acting constant gravity and perfectly elastic reflections from the walls of S, the moving point will pass arbitrarily close to every point of S.

Finally, the reader may observe some analogy between the motion of the reflected ray and the problem of the differential equation on the torus: let $x(t)$ satisfy the differential equation $x' = f(x, t)$ where f has period 1 in each variable [14]. The trajectory $(t, x(t))$ may be imagined to be a curve on the torus due to the periodicities of f. Here one also has the density theorem asserting that under suitable conditions the trajectory is either closed and periodic, or else dense on the torus surface; a theorem of the ergodic type has also been proved.

As an example of a three-dimensional problem whose solution uses reflection techniques, we consider a box shaped like a parallelopiped, and wrapped tightly with a thin smooth inextensible string in the form of a skew octagon, illustrated in Figure 6a.

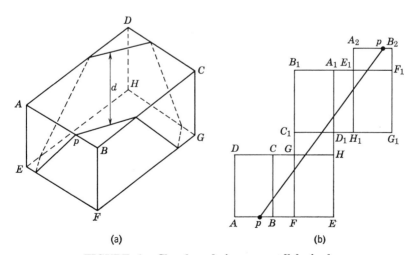

FIGURE 6.   Closed geodesic on a parallelopiped.

It is clear that the string can be moved along itself, remaining tight. We show that it can also be spread, so that the distance d changes, while still wrapping up the box tightly. For this we imagine the box cut open along its edges and spread out in the plane. Adding to the plane net certain reflections of faces in sides, we obtain the configuration of Figure 6b, in which the string appears as a straight segment. It is now clear that, within certain limits, the string can be moved parallel to itself, and each position corresponds to a different closed octagonal-shaped geodesic wrapping of the original parcel.

10. ROULETTES AND PEDALS

If a convex curve C rolls on the x-axis without slipping, then a point Q rigidly joined to C describes a curve C_1 called a roulette of C. To find the equation of C_1 in terms of that of C we refer to Figure 1. Let o be the origin of the Cartesian xy-system in Figure 1a; we suppose that C is tangent to the x-axis at o. Let $P(x, y)$ be a point on C and let s be the arc-length of C from o to P. Let Q have coordinates (a, b), let the X-axis be tangent to C at P and let the Y-axis be the distance s from P. Then the roulette C_1 traced out by Q is obtained as the trajectory of Q in the XY-system. Since $\tan \alpha = dy/dx$, we find, by taking projections in Figure 1b, the parametric equations of C_1:

$$X = s - \frac{(x - a) + (y - b)y'}{\sqrt{1 + y'^2}}, \qquad Y = \frac{(x - a)y' - (y - b)}{\sqrt{1 + y'^2}}$$

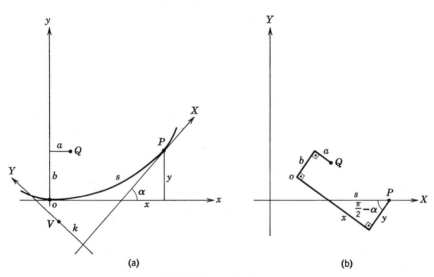

(a)   (b)

FIGURE 1. Roulettes.

if C is given by the equation $y = y(x)$. If C is given parametrically by $x = x(t)$, $y = y(t)$, then

$$X = s - \frac{(x - a)x' + (y - b)y'}{\sqrt{x'^2 + y'^2}}, \qquad Y = \frac{(x - a)y' - (y - b)x'}{\sqrt{x'^2 + y'^2}}.$$

If C is given by its natural equation, of the form $\alpha = \alpha(s)$, then its Cartesian coordinates are

$$x = \int_0^s \cos \alpha \, ds, \qquad y = \int_0^s \sin \alpha \, ds$$

and the parametric equations of the roulette C_1 are

$$X = s - (x - a) \cos \alpha - (y - b) \sin \alpha, \qquad Y = (x - a) \sin \alpha - (y - b) \cos \alpha.$$

For instance, when C is the circle of radius r, we have $s = r\alpha$ so that $x = r \sin \alpha$, $y = r(1 - \cos \alpha)$; putting $a = b = 0$ we obtain the parametric equations of the cycloid: $X = r(\alpha - \sin \alpha)$, $Y = r(1 - \cos \alpha)$.

The reader may wish to show that when the parabola $y = x^2/2p$ rolls without slipping on the x-axis, its focus (initially at $(0, p/2)$ describes the catenary $y = (p/2)\cosh(2x/p)$.

As a dual problem to that of a roulette, we consider a point V with fixed coordinates $(0, k)$ in the XY-system (see Figure 1a) and we consider its trajectory in the xy-system. We let u, v be the coordinates of V in the xy-system; taking projections in Figure 1a we have

$$u = x - s \cos \alpha - k \sin \alpha, \qquad v = y - s \sin \alpha + k \cos \alpha.$$

For instance, let C be the catenary $y = \cosh x - 1$ and let $k = -1$; we find then $y' = \tan \alpha = \sinh x$, $s' = \cosh x$, $s = \sinh$, $\sin \alpha = \tanh x$, $\cos \alpha = \operatorname{sech} x$ and hence $v = -1$. That is, V stays a constant distance below the y-axis. We may interpret this as follows.

Consider the track, shown in Figure 2b, consisting of repeated inverted segments of the catenary $y = \cosh x - 1$, $-\log (1 + \sqrt{2}) \leq x \leq \log (1 + \sqrt{2})$.

Suppose that a square wheel of side 2 rolls on this track without slipping, then the axle A of the wheel, which is at the height of the center of the rolling square, travels horizontally. In other words, we have here a track suitable for a train with square wheels.

In the case of compound roulettes, when a convex curve C rolls without slipping on a stationary convex curve C', the equations of the roulettes are somewhat harder to get and we proceed differently.

We consider the stationary curve C' with the natural equation $\beta = \beta(s)$ and the rolling curve with the natural equation $\alpha = \alpha(s)$ as shown in Figure 3a. Both curves are tangent to the x-axis at the origin; $P(x, y)$ is a point on C',

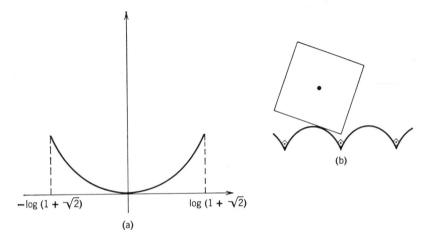

FIGURE 2.   Track for a square wheel.

$Q(u, v)$ is a point on C; the arc-lengths between O and Q on C and between O and P on C' are both s; $F(a, b)$ is a point rigidly connected to C. The rolling starts now and we consider the situation when Q has rolled over so as to co-incide with P, as shown in Figure 3b. To get the coordinates of F we refer to Figure 3c, and we find by applying projections onto the axes

$$X = x - (v - b)\sin(\alpha + \beta) - (u - a)\cos(\alpha + \beta)$$
$$Y = -y - (v - b)\cos(\alpha + \beta) + (u - a)\sin(\alpha + \beta),$$

(1)

where

$$x = \int_0^s \cos \alpha \, ds, \qquad y = \int_0^s \sin \alpha \, ds, \qquad u = \int_0^s \cos \beta \, ds, \qquad v = \int_0^s \sin \beta \, ds.$$

For instance, if the lower stationary curve is a circle of radius R and the upper rolling curve is a circle of radius r, and $a = b = 0$, we find the equation of the epicycloid which is the compound roulette here: $x = r \sin \alpha$, $y = r(1 - \cos \alpha)$, $u = R \sin \beta$, $v = R(1 - \cos \beta)$, $\alpha = s/r$, $\beta = s/R$, and

$$X = (r + R)\sin\alpha - R\sin(\alpha + \beta),$$
$$Y = -r + (r + R)\cos\alpha - R\cos(\alpha + \beta)$$

or

$$X = (r + R)\sin \frac{s}{r} - R \sin\left[\left(\frac{1}{r} + \frac{1}{R}\right)s\right],$$

$$Y = -r + (r + R)\cos \frac{s}{r} - R \cos\left[\left(\frac{1}{r} + \frac{1}{R}\right)s\right].$$

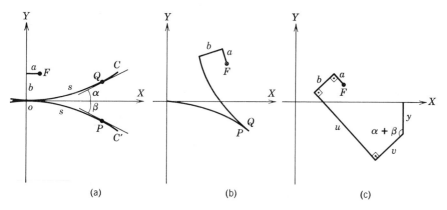

FIGURE 3.   Compound roulettes.

If r is changed to $-r$ in the above equations we get a hypocycloid instead of an epicycloid.

Formulas of the type (1) would be needed in such problems as that of the correct design of geared teeth. With reference to Figure 4 we ask: What are the correct profiles C and C' of the geared teeth so that the two wheels engage by rolling but there is no slipping and no grinding?

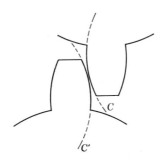

FIGURE 4.   Gear-teeth and compound roulettes.

Let C' be considered stationary and let C roll on C' without slipping. Let F be the center of the wheel carrying C and A be the center of the wheel carrying C'. Now the condition for correct design is that the distance between F and A stays constant; therefore the compound roulette path of F must be a circle about A.

Returning to the definition of simple roulettes, we consider again a convex curve C rolling without slipping on the x-axis as shown in Figure 5. Let P_1

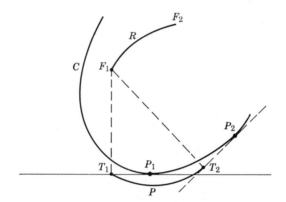

FIGURE 5.   Roulettes and pedals.

and P_2 be two points on C, and let C roll so that the point of contact with the x-axis changes from P_1 to P_2. Let a point rigidly joined to C describe then the arc R of the roulette from F_1 to F_2. Projecting the point F_1, fixed relative to C, onto every tangent to C between P_1 and P_2 gives us an arc P of the curve called the pedal of C relative to F_1.

It may be shown now, and the reader may wish to prove it, that the arcs P and R have the same length; further, the area under the arc R and above the base-line is twice the area of the sector subtended by P at F_1. For instance, when C is the circle $r = 2\,a \sin \theta$ and F_1 is the origin, the roulette is the cycloid $x = a(\theta - \sin \theta)$, $y = a(1 - \cos \theta)$, and the pedal is the cardioid $r = a(1 + \sin \theta)$. It follows that one arch of the cycloid has the same length as the complete cardioid, and encloses (against its base) twice the area of the cardioid.

11. INTEGRAL GEOMETRY

Integral geometry is concerned with assigning to certain geometrical events measures which remain invariant under certain groups of motions. For instance, we ask about the measure $m_1(C)$ of all straight lines in the plane which intersect an arc C, and we require that $m_1(C) = m_1(C_1)$ whenever C is congruent to C_1 under a rigid motion. Or, we ask about the measure $m_2(C)$ of all great circles on a sphere which cut a specified set C, asking additionally that $m_2(C) = m_2(C_1)$ if C and C_1 are congruent under a rotation of the sphere. Again, we ask for the measure $m_3(A, B)$ of all positions of an n-dimensional set A in E^n such that it intersects another such set B which is fixed; here we demand the invariance of $m_3(A, B)$ under all rigid motions applied to B. It will be noticed that the first problem is a special case of the third one, and that all three problems are similar except that in the second one we have a non-Euclidean space, the sphere, and its group of motions, instead of Euclidean cases. Since the measure such as $m_3(A, B)$ is defined in terms of motion of one set A relative to another set B, this measure is also called a kinematic measure, and a formula which expresses it in terms of A and B is also called a kinematic formula.

To motivate the insistence on the invariance of our measures we consider the so-called Bertrand paradox. We ask: What is the probability p that a random chord of a circle C exceeds in length the side of an equilateral triangle inscribed into C? By fixing one end of such a chord and observing that the other end is free to run over a third of C we would seem to get $p = 1/3$; by noticing that the midpoint m of the chord must be closer to the center o of C than to C we appear to have $p = 1/2$; finally, noting that m must be confined to the circle,. about o of half the radius, and so one quarter of the area, of C, we are led to $p = 1/4$. Each answer is a correct one but to a different problem, due to the looseness of the formulation of a "random chord." Nevertheless, the second answer $p = 1/2$ is "more correct" than others, precisely on the grounds of invariance.

To determine $m_1(C)$ we denote the straight line L by $L(\alpha, p)$ and we let its equation be

$$x \cos \alpha + y \sin \alpha - p = 0;$$

if

$$m_1(C) = \int dL = \iint\limits_{L(\alpha, p) \cap C \neq \varnothing} f(\alpha, p) \, d\alpha \, dp$$

then a calculation with the Jacobians shows that the invariance requirement forces $f(\alpha, p) = $ constant. This determines the measure up to a constant multiple. The indeterminacy of a constant multiple is due to the local compactness of the plane; in the compact case of a sphere it is possible to normalize so as to get rid of the multiple; there is also a connection here with absolute and with conditional probabilities. If the moving line is denoted by $L(a, b)$ with a, b its intercepts, then the invariance will give us

$$m_1(C) = \int dL = \text{const.} \iint\limits_{L(a, b) \cap C \neq \varnothing} ab(a^2 + b^2)^{-3/2} \, da \, db.$$

If C is a piecewise smooth rectifiable curve of length $l(C)$ we let its coordinates be $x(s)$, $y(s)$ in terms of the arc-length s, and we let θ be the angle which the line $L(s, \theta)$ intersecting C makes with the tangent to C at the point of intersection. The kinematic density is then

$$dL = |\sin \theta| \, ds \, d\theta, \qquad (1)$$

like all such densities it is always nonnegative. Integrating (1) over all lines L we find

$$\int n(L) \, dL = \int_0^{l(C)} \int_0^\pi |\sin \theta| \, d\theta \, ds = 2l(C).$$

Here $n(L)$ is the number of times that the line L cuts C; this is 0 if L is disjoint from C. If C happens to be a closed convex curve then $n = 0$ or $n = 2$ (for $n = 1$ only for the exceptional set—tangents to C—of measure 0) and so

$$\int_{C \cap L \neq \varnothing} dL = l(C).$$

If a curve C_1 is contained inside this closed convex C, we consider a line L intersecting C, and we ask for the number $n(L)$ of points in which this line also intersects C_1. We have then for the average $\bar{n}(L)$

$$\bar{n}(L) = \int_{L \cap C_1 \neq \varnothing} n(L) \, dL \bigg/ \int_{L \cap C \neq \varnothing} dL = 2l(C_1)/l(C). \qquad (2)$$

If C_1 is also a closed convex curve we have $L(C_1) < L(C)$ and we find the following probabilistic interpretation: if C and C_1 are closed convex curves, with C_1 inside C, then a line cutting C will also cut C_1 with the probability equal to the ratio of the lengths of the curves. If C_1 is arbitrary we have the following: some straight line must cut C_1 in at least $2l(C_1)/l(C)$ points. From the foregoing we can also deduce a very simple solution of the celebrated needle-problem due to Buffon (1777): a plane is lined with parallel straight lines one unit apart and a thin needle N of length $l < 1$ is thrown at random on the plane, what is the probability p that N intersects one of the lines? We imagine that N is rigidly connected to a circle C of radius $1/2$ centered at the middle of N. If the C-N configuration is thrown on the plane then exactly one line, say L. will intersect C (the probability of C fitting exactly in between two lines is clearly 0). We may therefore reverse the situation, so to speak, and regard C-N as stationary while a random line L cuts C. Now p is the probability that L also cuts N; therefore, from the preceding $p = 2l/\pi$.

By similar arguments one shows that the measure of all lines which cut a convex body K, of any finite dimension, is its surface area, and the measure of all planes which cut it is the integral M of mean curvature over the surface of K (assuming that K is sufficiently smooth).

A pair of independent straight lines L_1 and L_2 in the plane has the density

$$dL_1 \, dL_2 = dp_1 \, d\alpha_1 \, dp_2 \, d\alpha_2;$$

by the principle of evaluating one thing in two different ways we also express this density in terms of the coordinates x, y of the intersection of L_1 and L_2 and the angles β_1, β_2 which the lines make with the x-axis, getting

$$dL_1 \, dL_2 = |\sin (\beta_1 - \beta_2)| \, dx \, dy \, d\beta_1 \, d\beta_2$$

so that

$$dp_1 \, d\alpha_1 \, dp_2 \, d\alpha_2 = |\sin (\beta_1 - \beta_2)| \, dx \, dy \, d\beta_1 \, d\beta_2. \tag{3}$$

Let C be a closed convex curve of length $l(C)$ surrounding the region K of area $A(C)$; integrating (3) over all pairs of lines L_1, L_2 which cut C we find

$$l^2(C) = \iint\limits_{(x, y) \in K} |\sin(\beta_1 - \beta_2)| \, d\beta_1 \, d\beta_2 \, dx \, dy + \iint\limits_{(x, y) \notin K} |\sin(\beta_1 - \beta_2)| \, d\beta_1 \, d\beta_2 \, dx \, dy.$$

The first integral is $2\pi A(C)$; the second one, after some transforming, is found to be

$$2 \iint\limits_{(x, y) \notin K} (\omega - \sin \omega) \, dx \, dy$$

where $\omega = \omega(x, y)$ is the angle subtended by K at a point (x, y) outside K. Hence we have the Crofton formula

$$\iint\limits_{(x, y) \notin K} (\omega - \sin \omega) \, dx \, dy = 1/2 \, l^2(C) - \pi A(C).$$

As remarked by Bonnesen and Fenchel [6], this can be shown to be equivalent to the probabilistic statement: given a closed convex curve C as above and straight lines L_1, L_2 intersecting it, the probability that L_1 cuts L_2 inside C is $2\pi A(C)/l^2(C)$.

We consider next the measure $m_2(C)$ of all great circles E on a unit sphere S, which intersect a smooth (or piecewise smooth) arc C on S. The great circle, or equator, E is considered to be oriented and therefore a unique north pole corresponds to it; as the measure of a set of equators we take the area of the corresponding set of north poles. By another theorem of Crofton we have

$$m_2(C) = \int_{E \cap C \neq \varnothing} n(E) \, dE = 4l(C) \tag{4}$$

where $n(E)$ is the number of intersections of E and C, and $l(C)$ is the length of C. The same formula (4) applies if C is merely a rectifiable arc, or a finite union of such arcs. From this we deduce the solution of the following problem: What is the shortest length of thin inextensible smooth string which will tie up the unit sphere S with a net C on S so that S cannot fall out through any mesh of the net? This net C consists of a finite number of great-circle arcs (the geodesics on S) which meet at certain points. We show first that any equator E cuts C in at least three points: $n(E) \geq 3$. For if $n(E) = 0$, 1, or 2, then some simple geometry shows that by sliding a part of C over S we can remove S from the net (all one needs is to perform some rotations of certain parts of C). Hence

$$4l(C) \geq 3 \int dE = 12\pi$$

since the integral itself is just the area of the sphere. Thus $l(C) \geq 3\pi$. Next, for any $\varepsilon > 0$ we can construct an admissible net C of length $<3\pi + \varepsilon$ by taking three meridians of S making, say, $120°$ angles at the poles, and modifying this over a sufficiently small neighborhood of one pole of S. The reader may wish to complete this by showing that no net of length 3π will do; this amounts to showing that not only is $n(E) \geq 3$ but actually $n(E) > 3$ for a set of equators of positive measure.

We finish this section with proving by the methods of integral geometry the Bonnesen estimate of the isoperimetric deficit, mentioned in Section 7 of this chapter: if C is a closed convex curve of length $l(C)$ enclosing area $A(C)$, and if $R(C)$ and $r(C)$ are the radii of the smallest circle containing C and the greatest one contained in C, then

$$l^2(C) - 4\pi A(C) \geq \pi^2 [R(C) - r(C)]^2. \tag{5}$$

We shall need (special cases of) two of the principal formulas of integral geometry, due respectively to Poincaré and to Blaschke; both refer to the measure of the type $m_3(A, B)$ mentioned at the beginning of this section. Let A and B be two plane convex domains and $n(A, B)$ the number of times the boundary of the moving domain B cuts that of the stationary domain A, then the Poincaré formula is

$$\int_{B \cap A \neq \varnothing} n(A, B) \, dB = 4l(A)l(B) \tag{6}$$

and the Blaschke formula is

$$\int_{B \cap A \neq \varnothing} dB = 2\pi[A(A) + A(B)] + l(A)l(B), \tag{7}$$

here $A(X)$ refers to the area of the domain X and $l(X)$ to the length of its perimeter. We apply (6) and (7) to the case when the stationary domain A is that enclosed by C and the moving domain B is the circular disk of radius r, where

$$r(C) \leq r \leq R(C). \tag{8}$$

We get in this way

$$\int_{B \cap A \neq \varnothing} n(A, B) \, dB = 8\pi r l(C) \qquad (9)$$

$$\int_{B \cap A \neq \varnothing} dB = 2\pi A(C) + 2\pi^2 r^2 + 2\pi r l(C). \qquad (10)$$

Let M_n be the measure of all positions of B, such that $A \cap B \neq \varnothing$ and the boundaries of A and B intersect in exactly n points. Since both boundaries are closed convex curves, we have $M_n = 0$ for n odd; also, $M_o = 0$ by (8) and the extremal property of $r(C)$ and $R(C)$. Now (9) and (10) become

$$\sum_{j=1}^{\infty} 2j M_{2j} = 8\pi r l(C)$$

$$\sum_{j=1}^{\infty} M_{2j} = 2\pi A(C) + 2\pi^2 r^2 + 2\pi r l(C);$$

multiplying the first equation by $1/2$, the second one by 1, and subtracting, we have

$$\sum_{j=1}^{\infty} (j - 1) M_{2j} = 2\pi [r l(C) - A(C) - \pi r^2]$$

and since $M_n \geq 0$ we conclude that

$$r l(C) - A(C) - \pi r^2 \geq 0. \qquad (11)$$

Let the roots of the quadratic equation in r

$$\pi r^2 - l(C) r + A(C) = 0$$

be r_1 and r_2, with $r_1 \leq r_2$; then by (11) we must have

$$r_1 \leq r(C) \leq R(C) \leq r_2$$

so that

$$R(C) - r(C) \leq r_2 - r_1 = [l^2(C) - 4\pi A(C)]^{1/2} \pi^{-1}$$

and squaring this we get the Bonnesen inequality (5). For general references to integral geometry, see [6], [12], and [65].

12. CONVEXITY

We shall limit ourselves to some few very simple and special remarks on this subject, and to one problem. The definition of a convex set S is simple: together with any two points S also contains the straight segment joining them. Otherwise put: S is convex if every point of it is *visible via* S from any other point. This visibility aspect occasionally serves as a basis of a proof. For instance, let K be a closed convex n-dimensional set, of diameter 1 and with a smooth boundary ∂K (this means that there is a unique tangent plane to K through every point of ∂K). We wish to show that K is a union of $n + 1$ sets K_i of diameters <1. To do this, it is enough to show that ∂K is a union of $n + 1$ such sets V_i, of diameter <1; for then we can take any point o in the interior of K and form K_i as the conical set with o as vertex and V_i as curved base. To obtain the sets V_i we inscribe K into the interior of a simplex with vertices $v_1, \ldots, v_{n+1}$, and we let V_i be the set of all points of ∂K visible from v_i without obstruction by K. Each V_i has diameter <1 and each point of ∂K is in some V_i (here we use the useful diameter principle: if the diameter of a convex set X is d then $d = |xy|$ where x and y are extreme points of ∂X and there exist then support planes to X through x and y, orthogonal to xy). Briefly: the north and the south poles of any *smooth* object cannot be seen simultaneously from any finite distance.

Another intuitive physical aspect of convexity is seen in the problem of three double normals: given a three-dimensional convex body K we consider a support plane P at a point $x \in \partial K$, and we let the corresponding inward-drawn normal N to K at x cut ∂K again at y; if N is also normal to K at y it is called a double normal; now the problem is to show that K has at least three distinct double normals. We can express the problem, and gain some clues to its solution, thus: a convex shape K can be held in a parallel-jaw vise in at least three different positions. It may be observed that two such positions present no difficulty: they correspond to the maximal and the minimal separation of the jaws. We can also consider what happens to the point of contact of K with each jaw and to its projection onto the plane of the opposite jaw (of course, for some positions there may be a line or even an area of contact).

Since one double-normal configuration corresponds to the maximum of the jaw-separation, and the second one to its minimum, the reader may wonder about the *third* configuration. If d is the distance between the jaws of the vise, it can be shown that the third double-normal is associated with the saddle-point of d. The reader may wish to derive the existence of three double normals from the elements of the Morse theory of critical points on manifolds [50].

To distinguish between convex bodies and other convex sets, we consider a convex set X in the Euclidean n-space, and we observe that for many purposes X may be assumed to be closed; now we attempt to classify closed convex sets X. First, we ask if X contains a complete straight line L; if it does then it is a cylinder. For, taking a section S of X by a plane P orthogonal to L, we let p and q be two points on L on the opposite sides of P. The finite cones with base S and vertices p and q are in X; letting p and q move on L away from P proves the assertion. This process may be repeated and eventually we represent X as the Cartesian product of a complete $(n - k)$-dimensional plane

F, and a convex set X_1 in the k-dimensional space complementary to F. We have therefore reduced our classification problem to classifying closed convex sets X in k dimensions, which contain no complete straight line. If such a set is bounded, and k-dimensional, it is called a convex body, though it is only when X is referred to its ambient k-dimensional space that it is called a convex body. For instance, the closed unit disk in the plane is a convex body in the plane, and it has then a nonempty interior; when considered as a subset of three-dimensional space, it is not a convex body, and here its interior is empty.

Finally, we have closed convex k-dimensional sets which contain no lines but are unbounded. Let X be such a set, x a point in it, and consider all the unit vectors u which start at x; their end-points range over the unit sphere S about x. For each u we define $f(u)$ to be the length of that part of the straight ray $R(u)$ from x along u, which is in X. Since $f(u)$ is unbounded it follows that there exists u such that the whole ray $R(u)$ is in X. Let $A(X)$, the aperture of X, be the set of all u such that $R(u) \subset X$. $A(X)$ is independent of the base point x, and we may describe it as the set of all directions of unobstructed visibility in X. $A(X)$ is a closed subset of S and it is spherically convex; with every two points u and v it also contains the shorter arc of the great circle through u and v. (Note here: by hypothesis X does not contain a whole line, hence $A(X)$ does not contain a pair of opposite points.)

We next define the inner aperture $I(X)$ of X. This is a subset of $A(X)$, consisting of all u such that if we take $R(u)$ and a point v on $R(u)$, and let $r(v)$ be the distance from v to the boundary of X, then $r(v) \to \infty$ as v recedes on $R(u)$ away from x. The sets $A(X)$ and $I(X)$ are of some possible importance; the reader may consider two convex sets X and Y, and attempt to use $A(X)$, $I(X)$, $A(Y)$, $I(Y)$ to express the necessary and sufficient condition on X and Y so that a translate of X be strictly separable by a plane from Y.

Many problems can be raised in connection with the apertures; for instance, it is easy to show that

$$\text{interior } A(X) \subseteq I(X) \subseteq A(X)$$

but it seems to be unknown which sets lying between the two extremes above are admissible, i.e., actually occur as inner apertures.

While the metric concepts of length, area, volume, etc., present considerable difficulty for general sets, the definitions and theory are much simpler for convex sets. For instance, every closed convex curve C has a length and encloses an area, as can be seen by bracketing C between two sequences of inscribed and circumscribed polygons. However, the existence of length is intuitively braced by observing that C can roll without slipping on a straight line L, and its length is the distance between any two successive positions on L of a fixed point of C. Another situation involving a similar kinematic picture is the following: given two closed convex curves C_1 and C_2, with C_2 enclosing C_1, we have length (C_1) < length (C_2). For, we can bring C_1 inside C_2, to a contact with C_2 and then roll it inside C_2. We observe that the point of contact may "jump" but it is intuitively plausible that C_1 makes a full turn before going all the way round C_2. The apparent difficulty is that C_1 may get stuck inside C_2; this would occur for a long thin ellipse inside a similar ellipse. We overcome this by reflecting C_1 to the outside of C_2, if and where necessary, by a mirror-reflection of C_1 in a tangent common to both C_1 and C_2. A formal proof may be obtained by replacing C_1 and C_2 with sufficiently well-approximating convex polygons, and then projecting each side of the inner polygon onto the outer polygon.

In dealing with convex bodies we often need the existence of limits. For convex bodies there is a Bolzano–Weierstrass theorem known here as the Blaschke selection principle. It asserts that a bounded sequence of convex bodies has a convergent subsequence (it is necessary to bound both from outside and from inside, otherwise a sequence of n-dimensional convex bodies could collapse in the limit onto a lower-dimensional object). For these reasons of convenience (existence of lengths, areas, volumes, limits) convex bodies have been generalized in various ways so as to preserve these nice properties. One such generalization is suggested by the visibility approach. Instead of asking that every point of K should be visible via K from *every* other point of K, we demand more modestly that every point of K should be visible via K from all points of a subset $H(K)$ of K. We call $H(K)$ the hub of K; $H(K)$ is always convex. We observe that if K is the domain bounded by a simple plane sufficiently smooth closed curve, then $H(K)$ is simply determined by K and the tangents to K drawn at the points of inflection of K. This is shown on an example in Figure 1. With suitable generalization this extends to n dimensions. We suppose then that the surface bounding K is sufficiently smooth and we replace the tangent lines at the points of inflection by tangent planes at the parabolic points [45].

A not unnatural condition on $H(K)$ is that it should not be too small or flat. We call K a generalized convex body if $H(K)$ contains a ball B of the same dimension as K. In other words, we demand that $H(K)$ should also be a convex body. It turns out that lengths, areas, volumes, and limits exist for generalized convex bodies just as they do for convex bodies. In proving the equivalent of Blaschke's selection principle for a sequence K_1, K_2, ... of generalized convex bodies we assume that the hubs $H(K_1)$, $H(K_2)$, ... contain a *common* ball B; this implies a certain uniformity property for the sequence K_1, K_2, ... It turns out that any generalized convex body K is simply given by a generalization of polar coordinates. We let o be the center of a ball B contained in $H(K)$ and for any point s in the boundary S of B we produce os beyond s till it cuts the boundary of K, in s_1, say. Now we express the length

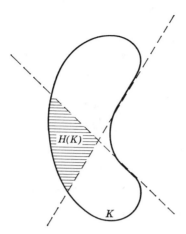

FIGURE 1. The hub of a plane set.

$|os_1|$ as a function of s on S. K is completely given by this function $f(s)$ and the reader may wish to show that f satisfies the Lipschitz condition. Conversely, any positive function $f(s)$ on S, which satisfies a Lipschitz condition, determines a generalized convex body. The reader may wish to show that there always exists a positive constant a, such that the function $f(s) + a$ determines a *convex* body.

Next, we present the use of one of the principal results in convexity, the Brunn–Minkowski theorem, as it occurred in a biological problem. A microscopic object X (a cell) was examined by taking n equispaced parallel sections and looking at them under a microscope. The areas $A_1, \ldots, A_n$ of the sections could thus be found and the question arose: knowing $A_1, \ldots, A_n$ what can one say about the volume $V(X)$? Without some further assumption, the answer is clearly: nothing. It was then assumed that X was convex (we needn't justify this here.) Now the problem is as follows: given a convex solid X contained between two parallel supporting planes P_0 and P_{n+1} one unit apart, given the sequence $P_0, P_1, \ldots, P_{n+1}$ of equispaced parallel planes, and given the area A_i of the section of X by P_i, $i = 1, 2, \ldots, n$, what absolute bounds $V_{\min}$ and $V_{\max}$ can be assigned to the volume $V(X)$?

The difficulty of this problem is that there is too much to vary and perturb. We obtain the crucial reduction from a three-dimensional to a two-dimensional problem by symmetrizing X with respect to the line L orthogonal to the planes P_i. The result of this operation is a solid Y of *revolution* about L uniquely determined by the following condition: any plane orthogonal to L cuts both X and Y in sections of the same area. It follows by the Cavalieri principle that X and Y have equal volumes. Next, the Brunn–Minkowski theorem asserts that Y too is convex. Therefore, if Y is obtained by rotating about the x-axis the graph of $y = \phi(x)$, $0 \le x \le 1$, it follows that ϕ is a concave function (its arc never lies below the corresponding chord). Now the problem becomes this: to find two nonnegative continuous concave functions $f(x)$ and $F(x)$, defined for $0 \le x \le 1$, which satisfy

$$f\left(\frac{j}{n+1}\right) = F\left(\frac{j}{n+1}\right) = (A_j/\pi)^{1/2}, \qquad j = 1, \ldots, n$$

and are subject to the conditions

$$V_{\min} = \pi \int_0^1 f^2(x)\, dx \quad \text{is smallest possible,}$$

$$V_{\max} = \pi \int_0^1 F^2(x)\, dx \quad \text{is largest possible.}$$

It is not hard to show that $f(x)$ is piecewise linear, $f(0) = f(1) = 0$, and $f(x)$ is linear over each interval

$$\left(\frac{j}{n+1}, \frac{j+1}{n+1}\right), \qquad j = 0, 1, \ldots, n.$$

This allows us to calculate $V_{\min}$ explicitly (using the prismoidal formula, for instance). $F(x)$ and $V_{\max}$ are harder to get at but it is not hard to show that $F(x)$ is also piecewise linear; however, the slopes of its linear segments satisfy a rather messy set of conditions. But it is relatively easy to find a reasonable estimate of the form

$$V_{\min} \le V_{\max} \le V_{\min} + A$$

where the quantity A depends on n, on the maximal section area A_k (assuming $1 < k < n$), and on A_{k-1} and A_{k+1}.

The following important further property of the symmetrization may be noticed: if Y is the symmetrization of X, not only is $\text{vol}(X) = \text{vol}(Y)$ but also $\text{area}(X) \le \text{area}(Y)$ and the equality holds only if $X = Y$ (i.e., if X is already symmetric in that particular direction). This was used by Schwarz in his solution of the isoperimetric problem of the sphere: of all closed surfaces S of fixed surface area the sphere encloses the largest volume. In fact, for the case of convex S we obtain the solution almost immediately: such maximal convex S must be unchanged by symmetrization in any direction; hence it is a sphere. However, an important distinction between the isoperimetric problems of the circle and of the sphere must be observed. Replacing a plane closed curve by the boundary of its convex hull cannot increase the length, whereas replacing a closed surface by the boundary of its convex hull may very well increase the area. Thus the reduction to a convex case is immediate for the circle problem and not at all so for the sphere problem (this is one of the reasons why the latter is so considerably harder).

13. CURVES IN *n* DIMENSIONS

As an exercise in vector algebra and elementary properties of matrices we derive here the Frenet–Serret formulas for curves. A curve C in the n-dimensional Euclidean space is conveniently given parametrically by one vector equation $x = x(t)$; t is a scalar variable, usually the length of C measured from some fixed point, and the Cartesian coordinates are then $(x_1(t), x_2(t), \ldots, x_n(t))$. We suppose that the functions $x_i(t)$ are sufficiently differentiable, we fix a point $p = x(t_0)$ on C, and we determine the successive osculating flats to C at p. The first one is the tangent line at p, or the line of closest contact with C at p, the second osculating flat is the two-dimensional plane through p of closest possible contact with C, and so on. In general, to determine the kth osculating flat T_k we take $k + 1$ distinct values of $t : t_0, t_1, t_2, \ldots t_k$, and we consider the lowest-dimensional flat containing $x(t_0), x(t_1), \ldots, x(t_k)$; its limiting position as $t_1, t_2, \ldots, t_k \to t_0$ is T_k. The point $p = x(t_0)$ is called regular if T_k is k-dimensional, or equivalently, if the n vectors $x'(t_0), x''(t_0), \ldots, x^{(n)}(t_0)$ are linearly independent. Each T_k contains then the previous ones and T_n is the whole ambient space. By the usual orthonormalizing process we pass from $x'(t_0), x''(t_0), \ldots, x^{(n)}(t_0)$ to the corresponding orthonormal n-tuple of vectors $a_1(t_0), a_2(t_0), \ldots, a_n(t_0)$, with the first k vectors of each n-tuple spanning the same space, namely the flat T_k. For $n = 3$ these a's are called the tangent, the principal normal, and the binormal respectively. The plane through a_2 and a_3 is called the normal plane, the one through a_1 and a_3 the rectifying plane, and the one through a_1 and a_2 the osculating plane.

Since $a_1, \ldots, a_n$ form a basis it follows that

$$a_i' = \sum_{k=1}^{n} \rho_{ki} a_k, \qquad i = 1, \ldots, n; \tag{1}$$

let R be the matrix (ρ_{ki}). By the orthonormality of the a_i's we have

$$a_i \cdot a_j = \delta_{ij}, \qquad \delta_{ij} = 0 \quad \text{if} \quad i \neq j, \qquad \delta_{ii} = 1.$$

Hence by differentiating we find that

$$a_i \cdot a_j' + a_i' \cdot a_j = 0$$

so that

$$\rho_{ki} + \rho_{ik} = 0$$

or R is skew-symmetric. Further, from the description of the successive osculating flats and their connection with the vectors a_k it follows that a_k is linearly expressible by the derivatives $x'(t_0), \ldots, x^{(k)}(t_0)$ and no higher, and therefore a_k' is expressible by the derivatives $x'(t_0), \ldots, x^{(k+1)}(t_0)$ and no higher. Equivalently, a_k' is a linear combination of $a_1, a_2, \ldots, a_{k+1}$. Thus the matrix R has all elements 0 above the first superdiagonal; being also skew-symmetric, it is of the form

$$
\begin{pmatrix}
0 & \rho_1 & 0 & 0 & \cdots \\
-\rho_1 & 0 & \rho_2 & 0 & \cdots \\
0 & -\rho_2 & 0 & \rho_3 & \cdots \\
0 & 0 & -\rho_3 & 0 & \cdots \\
\cdots & \cdots & \cdots & \cdots & \cdots \\
\cdots & \cdots & \cdots & \cdots & \cdots
\end{pmatrix}
$$

Therefore the equations (1) become the Frenet–Serret formulas

$$
\begin{aligned}
a_1' &= & \rho_1 a_2 & & \\
a_2' &= -\rho_1 a_1 & & +\rho_2 a_3 & \\
a_3' &= & -\rho_2 a_2 & & +\rho_3 a_4 \\
& \cdots & & & \\
a_{n-1}' &= & -\rho_{n-2} a_{n-2} & & +\rho_{n-1} a_n \\
a_n' &= & -\rho_{n-1} a_{n-1}. & &
\end{aligned}
$$

It may be verified that the quantities $\rho_i(t_0)$ are not invariant when we change variables $(t = f(\tau))$ but that their ratios are. When t is the arc-length on C the quantities $\rho_1(t_0), \ldots, \rho_{n-1}(t_0)$ are called the curvatures of C at $p(t_0)$. For $n = 3$, $\rho_1(t_0)$ is also called the curvature and $\rho_2(t_0)$ the torsion.

The kth osculating flat is spanned by the vectors $a_1, \ldots, a_k$; the kth curvature $\rho_k(t_0)$ gives us a measure of how fast the next vector a_{k+1} departs from T_k. Alternatively, $\rho_k(t_0)$ tells us how fast the flat T_k twists as we move along the curve. The reciprocals $1/\rho_k(t_0)$ are known as the radii of curvature.

The reader may wish to show that if $p(t_0)$ is taken as the origin and $a_1(t_0)$, $\ldots, a_n(t_0)$ are the unit vectors, then the parametric equations of C referred to the new coordinate system are

$$
x_k(t) = c_k t^k + O(t^{k+1}), \qquad k = 1, \ldots, n; \tag{2}
$$

further, the reader may wish to express the quantities c_k in terms of the curvatures $\rho_k(t_0)$.

Equations (2) have some connection with a possible extension of convexity. A plane curve K is convex if and only if a straight line in its plane which cuts K, cuts it the "right" and "minimal" number of times: twice (we exclude tangent lines). This suggests an extension to curves in n-space: such a curve is n-convex if and only if a k-dimensional flat which cuts it, cuts it the "right" and "minimal" number of times: $k + 1$ times. We amend it to: the k-flat cuts it no more than $k + 1$ times. The reader may wish to show that if we drop the O-terms in (2) and the c_k do not vanish, one gets an n-convex curve. Further, the reader may wish to re-formulate it in some such way as the following: an n-curve which is regular is locally n-convex.

2

ITERATION

1. PRELIMINARIES

A function $f(x)$ is iterated by using its value as the new argument: $f(x) \to f(f(x))$. Starting with x we have the sequence $x, f(x), f(f(x)), \ldots$ which we call 0th, 1st, 2nd, $\ldots$ iterates of f and which we denote by $f_0(x), f_1(x), f_2(x), \ldots$ or briefly by $f_0, f_1, f_2, \ldots$. The index rule $f_n(f_m(x)) = f_m(f_n(x)) = f_{m+n}(x)$ is easily verified. The successive iterates $x_0, x_1 = f(x_0), x_2 = f(x_1), \ldots$ are simply obtained from the graphs of $y = f(x)$ and $y = x$ as shown in Figure 1. If the sequence $x_0, x_1, x_2, \ldots$ of iterates converges, then the limit must be a fixed point of $f(x)$, that is, a root of the equation $x = f(x)$. This serves as a basis for solving the equation $F(x) = 0$; we write it as $x = x + F(x)$ and we iterate the function $f(x) = x + F(x)$.

Expressing $f_n(x)$ explicitly can be done for a few simple functions:

$$\text{if} \quad f(x) = x^a \qquad f_n(x) = x^{a^n}$$

$$\text{if} \quad f(x) = cx \qquad f_n(x) = c^n x,$$

$$\text{if} \quad f(x) = ax + b \qquad f_n(x) = a^n x + b\,\frac{a^n - 1}{a - 1}.$$

The last example generalizes slightly: if $f(x) = a_1 x + b_1$, $g(x) = a_2 x + b_2$, $\ldots$, $h(x) = a_n x + b_n$ then

$$h(\ldots g(f(x))\ldots) = \prod_{i=1}^{n} a_i x + \sum_{j=1}^{n} \left(b_j \prod_{i=j+1}^{n} a_i \right).$$

EXAMPLE. Generalizing a well-known puzzle we ask: n men and m monkeys gather coconuts all day and then fall asleep; the first man wakes up, sets apart p coconuts for each monkey and one nth of what remains for himself and goes back to sleep; shortly after, the second man wakes up and repeats

51

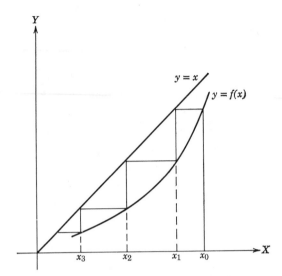

FIGURE 1.   Staircase diagram for iterates.

the same procedure on the remaining coconuts; then the third man, fourth, and so on; in the morning all men wake up, give p coconuts to each monkey and one nth of what remains to each man; assuming that all divisions came out in integers what was the minimum number of coconuts originally gathered?

Let that be x and let $f(x) = [(n-1)/n](x - mp)$, if $f_j(x)$ is the jth iterate then $f_j(x)$ coconuts remain just after the jth man has finished. In particular, $J_n(x)$ coconuts remain in the morning and this number must be of the form $nk + mp$. Using the explicit form of the nth iterate of f, we have therefore

$$x = \frac{n^{n+1}}{(n-1)^n}(k + mp) - mp(n-1);$$

since this is a positive integer, k is of the form

$$k = t(n-1)^n - mp$$

so that

$$x = tn^{n+1} - mp(n-1).$$

Therefore the answer is: x as given by the above, where t is the smallest integer which makes x positive.

2. EXPLICIT ITERATION AND CONJUGACY

The general problem of finding an explicit formula for $f_n(x)$ is hard. The most useful method here is that of conjugate functions. Denoting throughout the composition $f(F(x))$ by fF, let us suppose that

$$\phi f = g\phi \qquad (\text{or} \quad f\psi = \psi g)$$

where g is such that the iterates g_n can be determined explicitly. Then f_n can also be found for we have $f = \phi^{-1}g\phi$ (or $= \psi g \psi^{-1}$) so that

$$f_2 = \phi^{-1}g\phi\phi^{-1}g\phi = \phi^{-1}g_2\,\phi \qquad (\text{or} \quad = \psi g_2 \psi^{-1})$$
$$f_3 = \phi^{-1}g_3\,\phi \qquad\qquad\qquad (\text{or} \quad = \psi g_3 \psi^{-1})$$

and generally

$$f_n = \phi^{-1}g_n\,\phi \qquad\qquad\qquad (\text{or} \quad = \psi g_n \psi^{-1}).$$

Since g_n is known, so is f_n; in the special case when $g(x) = x + 1$, $\phi(x)$ is known as the Abel function, when $g(x) = cx (c \neq c^2)$, $\phi(x)$ is known as the Schroeder function. In the general case we call f and g conjugate under ϕ (or ψ).

Let

$$f(x) = a_1 x + a_2 x^2 + \cdots, \qquad 0 < a_1 < 1,$$

be a power series with positive radius of convergence, let x_0 be an arbitrary sufficiently small positive number, and let $x_n = f_n(x_0)$ be the nth iterate. A Schroeder function with $c = a_1$ may be obtained as follows. We have

$$\frac{x_{k+1}}{a_1 x_k} = 1 + \frac{a_2}{a_1}x_k + \frac{a_3}{a_2}x_k^2 + \cdots = 1 + r_k,$$

say. We use the multiplicative form of the telescoping cancellation: write $0, 1, \ldots, n-1$ for k in the above, and multiply the n equations; this yields

$$\frac{x_n}{a_1^{\,n}x_0} = \prod_{k=0}^{n-1}(1 + r_k).$$

By the standard convergence criteria for infinite products the limit as $n \to \infty$ exists, therefore

$$\lim \frac{x_n}{a_1^{\,n}} = \psi(x_0).$$

This is indeed a Schroeder function, for

$$\psi(f(x_0)) = \psi(x_1) = \lim_{n \to \infty} \frac{x_{n+1}}{a_1^n} = a_1 \lim_{n \to \infty} \frac{x_{n+1}}{a_1^{n+1}} = a_1 \psi(x_0)$$

so that $\psi f = a_1 \psi$.

EXAMPLES

(a) $f(x) = 2x^2 - 1$, $\phi(x) = \cos x$, $g(x) = 2x$ (thus ϕ is the Schroeder function and $c = 2$). Here $f\phi = \phi g$ so that $f = \phi g \phi^{-1}$ and hence

$$f_n(x) = \cos(2^n \arccos x).$$

(b) $f(x) = 4x(1 + x)$, $\phi(x) = \sinh^2 x$, $g(x) = 2x$. Here

$$f_n(x) = \sinh^2(2^n \arg \sinh \sqrt{x}).$$

(c) $f(x) = 4x(1 - x)$, same as above but $\sin^2 x$ replaces $\sinh^2 x$.
(d) $f(x) = x^2 - 2$. We observe that $f(x + 1/x) = x^2 + x^{-2}$; hence we let

$$\phi(x) = x + \frac{1}{x}, \qquad s(x) = x^2$$

to obtain $f = \phi s \phi^{-1}$. To compute ϕ^{-1} we solve a quadratic equation and we have

$$\phi^{-1}(x) = \frac{x \pm \sqrt{x^2 - 4}}{2};$$

the problem of the ambiguity of sign does not really arise since the two values are each other's reciprocals, and we have $f_n = \phi s_n \phi^{-1}$, or

$$f_n(x) = \left(\frac{x + \sqrt{x^2 - 4}}{2} \right)^{2^n} + \left(\frac{x - \sqrt{x^2 - 4}}{2} \right)^{2^n}.$$

Expanding by the binomial theorem and canceling, we find the polynomial form of the answer:

$$f_n(x) = 2^{1 - 2^n} \sum_{k=0}^{2^n - 1} \binom{2^n}{k} x^{2^n - 2k} (x^2 - 4)^k.$$

(e) Suppose that functions ϕ and g have been found so that $f = \phi g \phi^{-1}$ and that g is explicitly iterable. Let $L(x) = ax + b$, $L^{-1}(x) = (x - b)/a$, then both f and $L^{-1}fL$ are also explicitly iterable for

$$L^{-1}fL = L^{-1}\phi g \phi^{-1} L = (L^{-1}\phi) g (L^{-1}\phi)^{-1}.$$

For instance, let $f(x) = x^2 - 2$ as in the previous example, then

$$L^{-1}fL = ax^2 + 2bx + \frac{b^2 - b - 2}{a}.$$

Choosing, for instance, $b = 7$ and $a = 8$, we have

$$h(x) = L^{-1}fL(x) = 8x^2 + 14x + 5$$

and so

$$h_n(x) = \frac{f_n(8x + 7) - 7}{8}$$

where $f_n(x)$ is as in the previous example. Another useful form for L, besides the linear case, is the bilinear:

$$L(x) = \frac{ax + b}{cx + d}, \qquad L^{-1}(x) = \frac{b - dx}{cx - a}.$$

(f) Let A be an $m \times m$ matrix, to find A^n we determine first a nonsingular $m \times m$ matrix ϕ and an $m \times m$ matrix B so that $A = \phi B \phi^{-1}$; here A and B are called similar, rather than conjugate, under ϕ. We have then

$$A^n = \phi B^n \phi^{-1}$$

and if B happens to be diagonal:

$$B = \begin{pmatrix} b_1 & & & \\ & b_2 & & \\ & & \ddots & \\ & & & b_m \end{pmatrix}$$

then B^n is simply

$$B^n = \begin{pmatrix} b_1{}^n & & & \\ & b_2{}^n & & \\ & & \ddots & \\ & & & b_m{}^n \end{pmatrix}.$$

It is known from matrix theory that such a nonsingular ϕ and a diagonal B can be found if and only if A has all its eigenvalues distinct; $b_1, b_2, \ldots,$ b_m are then these eigenvalues. Even when B is not diagonal but consists of small square blocks:

$$B = \begin{pmatrix} B_1 & & & \\ & B_2 & & \\ & & \ddots & \\ & & & B_s \end{pmatrix}$$

with zeros elsewhere, the evaluation of powers A^n via the formula $A^n = \phi B^n \phi^{-1}$ may save considerably on the number of operations, as against the direct computation of A^n.

3. DUPLICATION AND TRIPLICATION

The reader may have noticed that in the previous examples for explicit iteration of a function $f(x)$ we meet often a Schroeder function $\phi(x)$ which is trigonometric or hyperbolic, with the corresponding constant $c = 2: f[\phi(x)] = \phi(2x)$. Briefly: f functions as the duplication for ϕ. Behind this fact lies a theorem of Weierstrass on algebraic addition formulas. A function $\phi(x)$ is said to possess an algebraic addition formula if there is a polynomial P in three variables, such that

$$P[\phi(x + y), \phi(x), \phi(y)] = 0 \tag{1}$$

identically for all x and y. The theorem of Weierstrass states that the only functions with algebraic addition formulas are elliptic functions and their special degenerate cases such as trigonometric, hyperbolic, and exponential functions. From the addition formula (1) we have the duplication formula

$$Q[\phi(2x), \phi(x)] = 0$$

with a polynomial Q. This suggests that $\phi(x)$ may be a Schroeder function for a suitable $f(x)$. Further, we notice the connection between duplication, triplication, etc., and the problem of explicit iteration.

As a further comment on triplication, we observe that the so-called irreducible case of a cubic equation can be solved by reference to the triplication formula of the cosine. We start by outlining briefly the solution of a cubic equation. The general cubic is taken in the form

$$x^3 + ax^2 + bx + c = 0 \tag{2}$$

and we put

$$y = x + a/3 \tag{3}$$

obtaining for y the reduced cubic

$$y^3 + py + q = 0 \tag{4}$$

where $p = b - a^2/3$, $q = 2a^3/27 - ab/3 + c$. Next, if $y = u + v$ then (4) becomes

$$u^3 + v^3 + (u + v)(3uv + p) + q = 0.$$

We put $3uv + p = 0$ so that $v = -p/3u$ and we obtain

$$u^6 + qu^3 - p^3/27 = 0$$

which is a quadratic in u^3. Hence u and v, and eventually x, can be determined.

Let us suppose that a, b, c in (2) are real and let us determine when (2) has three distinct real roots. Writing (2) as $f(x) = 0$ we find the following necessary and sufficient condition: the quadratic $f'(x) = 0$ must have two distinct real roots A and B which are such that $f(A)f(B) < 0$. No such simple criterion exists for higher degree equations.

When the cubic (2) has three distinct real roots, none of them rational, we have the irreducible case. It is known [75] that it is impossible then to express the roots by real radicals. In particular, we have the triplication formula

$$4 \cos^3 \alpha - 3 \cos \alpha - \cos 3\alpha = 0, \tag{5}$$

with $x = \cos \alpha$, $a = \cos 3\alpha$ we get the irreducible cubic

$$x^3 - \frac{3}{4} x - \frac{a}{4} = 0;$$

hence the trisection of an arbitrary angle is impossible by ruler and compass constructions. However, we can express the roots of a cubic in the irreducible case by trigonometric functions. First, using the above necessary and sufficient condition, we verify that the cubic (4) has three real distinct roots if and only if $27q^2 + 4p^3 < 0$. Let $y = 2(-p/3)^{1/2}v$, then (4) becomes

$$4v^3 - 3v - D = 0$$

with $D = 3q/[2p(-p/3)^{1/2}]$. Therefore $D^2 < 1$ and we can put $D = \cos 3\beta$; by reference to the triplication formula (5) we find that the roots of (4) are

$$2(-p/3)^{1/2} \cos(\beta + 2k\pi/3), \qquad k = 0, 1, 2,$$

This can also be written as

$$2(-p/3)^{1/2} \cos\left[\frac{2k\pi + \arccos D}{3}\right], \qquad k = 0, 1, 2,$$

showing similarity to the formulas for explicit iteration.

Quartic equations can also be solved by elementary algebraic means. We start with the reduced quartic

$$y^4 + My^2 + Ny + P = 0 \tag{6}$$

obtained by applying the analogue $y = x + a/4$ of (3) to the analogue of (2). We attempt to factorize (6) as

$$(y^2 + Ay + B)(y^2 - Ay + C) = 0 \tag{7}$$

and by comparing the coefficients in (6) and (7) we have

$$B + C = M - A^2, \qquad B - C = N/A, \qquad BC = P.$$

Since $(B + C)^2 + (B - C)^2 = 4BC$ we find that

$$(M - A^2)^2 - N^2/A^2 = 4P$$

which is a cubic in A^2. Hence A, B, C can be found and eventually we solve (6). On the analogy with the trigonometric solution of the cubic, the reader may wish to attempt a solution of the irreducible case of (6) by trigonometric means, using the quadruplication formula for $\cos x$.

It is known from the work of Ruffini, Abel, and Galois that, unlike the quadratics, cubics, and quartics, the general quintic equation cannot be solved by radicals [75]. However, as shown by Hermite [30], it is possible to solve the quintic by using the so-called elliptic modular functions. This solution proceeds in two parts: first, the quintic is reduced to simplest possible form; then we employ a relation between certain two elliptic modular functions which plays here the same role as the triplication formula (5) in the solution of the cubic. The reduction starts with the quintic analogue of (2):

$$x^5 + ax^4 + bx^3 + cx^2 + dx + f = 0 \tag{8}$$

but instead of the linear transformation (3) we use a higher-order one:

$$y = a_1 x^5 + b_1 x^4 + c_1 x^3 + d_1 x^2 + e_1 x + f_1. \tag{9}$$

This method is due to Tschirnhaus (letters to Leibniz, 1678–1683); when x is eliminated from (8) and (9) there results a quintic for y in which we have control over various coefficients by choosing $a_1, \ldots, f_1$ in (9) suitably. Using quadratic equations only we can put the resultant quintic in the form

$$y^5 + Ay^2 + By + C = 0.$$

If we wish to reduce it still further, to the form

$$y^5 + By + C = 0, \tag{10}$$

the elimination of x from (8) and (9) leads to a sextic equation. However, as shown by Bring, Jerrard, and Kronecker [37], this sextic can be solved by means of quadratics and cubics. Therefore we may assume that the quintic is in the form (10); we put $y = (-B)^{1/4}t$ and t satisfies the Hermite reduced form

$$t^5 - t - D = 0. \tag{11}$$

This completes the reduction process; we have obtained an equivalent quintic with a single variable parameter D.

Next, Hermite used a far-reaching generalization of the triplication formula (5). Before stating it, we observe that (5) could be restated as follows. Consider the two functions $f(x) = \cos x$ and $g(x) = \cos 3x$; then f has the period three times that of g, and the two functions are connected by a polynomial relation which is of degree 3 in f. For the elliptic modular functions the period interval $[0, 2\pi]$ or $[0, 2\pi/3]$ is replaced by the so-called fundamental region; a theorem [22] states that if two such functions f_1 and f_2 belonging to a certain family have fundamental regions S_1 and S_2, with S_1 being a union of k copies of S_2, then f_1 and f_2 are connected by a polynomial relation of degree k in f_1. Hermite found two elliptic modular functions $f(x)$ and $g(x)$ such that

$$f^5(x) - f(x) - g(x) = 0$$

identically; by reference to (11) he was able to express the roots of the quintic (11) in terms of f and g. For details and a numerical example see [17].

4. EXTENSION TO NONINTEGER ITERATION

The problem we consider here is one of a type which occurs often in mathematics: how to extend to a larger domain a quantity or an operation defined originally for integers alone. We ask: What, if anything, is the ath iterate f_a of f, for real a? If $f(x)$ is a $1 : 1$ function, there is the inverse f^{-1} and we define f_{-1} to be the inverse f^{-1}, with the preservation of all iteration rules. Thus we may limit ourselves to positive real iteration index a.

An answer is provided by the conjugate function method: if $f = \phi g \phi^{-1}$ and $g(x)$ is such a function as $x + 1$, cx, or x^s, then the ath iterate $g_a(x)$ is "naturally" given as $x + a$, $c^a x$, or x^{s^a}, and f_a may then be defined as $f_a = \phi g_a \phi^{-1}$. For instance, a solution of the functional equation

$$h(h(x)) = 2x^2 - 1 \tag{1}$$

is obtained with the help of a previous example as the half-order iterate $f_{1/2}$ of $f(x) = 2x^2 - 1$, given by

$$f_{1/2}(x) = \cos(\sqrt{2} \text{ arc cos } x).$$

But there are many other solutions of (1). It will turn out that half-order iteration is highly nonunique, so nonunique in fact that on a small interval a half-order iterate of a given function f may be defined almost arbitrarily. We explain this on the example of the function $f(x) = x^2$. The graph C of $y = x^2$ for $0 \leq x \leq 1$ is shown in Figure 1, together with the diagonal graph D of $y = x$, $0 \leq x \leq 1$.

Let R be any rectangle with the sides parallel to the coordinate axes, whose upper left corner lies on D and lower right on C. In R, let H be an *arbitrary* continuously increasing graph of a function, joining the other two corners of R. We now extend H by the "rectangular" continuation: let R move so that the lower left corner rides on H, the upper left one on D, and the lower right one on C. Then the upper right corner traces out a continuation of H to a larger domain of x's. We then repeat the same procedure on the enlarged graph, and we keep on repeating. In the same fashion, by moving the rectangle R to the left of its original position, we obtain a continuation of H

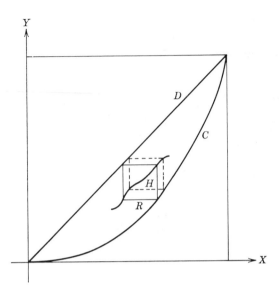

FIGURE 1. Half-order iteration of $f(x) = x^2$.

to the left. When the rectangular continuation is repeated indefinitely often in both directions, we obtain the graph of a function $h(x)$ for $0 < x < 1$, and now we define $h(0) = 0$, $h(1) = 1$. The completed graph H_1, the diagonal D, and the graph C of f are shown together in Figure 2. The graphs have the rectangle property: whenever any three vertices of a rectangle lie on their proper graphs as indicated, then so does the fourth vertex. But, as is obvious from the graph, this means that for any x, $0 \le x \le 1$, $h(h(x)) = f(x)$. Thus $h = f_{1/2}$ has been constructed which is essentially arbitrary over a subinterval.

The reader may ask himself whether there exists a way of eliminating the rather unpleasant superabundance of half-order iterates of f. One possibility might be to assume that f has some property P (of sufficient smoothness or regularity) and to demand that one and only one half-order iterate should exist, that also has the property P. Why choose the half-order iterates for the distinction of uniqueness? For if there exists a unique $f_{1/2}$, then there exist also unique $f_{1/4}, f_{1/8}, \ldots, f_{1/2^n}, \ldots$. Further, by composition we can generate then unique iterates of order $p/2^n$ with p an odd integer: $f_{p/2^n} = h_p$ where $h = f_{1/2^n}$. By further composition we generate unique iterates whose order is any dyadic fraction (i.e., $\sum_{i=0}^{n}(p_i/2^{n_i})$) and by a suitable passage to the limit we produce unique real-order iterates f_a, since the dyadic fractions are dense in the positive reals.

The reader who is acquainted with such matters may wish to compare the above paradigm with the proof of one of the crucial lemmas which led to the

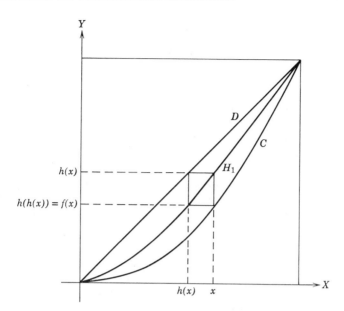

FIGURE 2.   The rectangle property.

solution of Hilbert's fifth problem. This problem is as follows. Let G be both a group and a topological space, satisfying two conditions: (a) as a topological space G obeys the Hausdorff separation axiom; (b) the algebraic and the topological structures of G are such that the algebraic operations, i.e., inverses and products, are continuous in the topology. Such a G is called a topological group; it is called locally Euclidean if a neighborhood V of the identity e of G is homeomorphic to an Euclidean space of some dimension n. There is a neighborhood $U \subseteq V$, such that if x and y are in U then x^{-1} and $z = xy$ are in V. It follows then that any homeomorphism ϕ of V with the Euclidean n-space allows us to introduce coordinates into V and we have then

$$x = (x_1, \ldots, x_n), \quad y = (y_1, \ldots, y_n), \quad z = (z_1, \ldots, z_n), \quad x^{-1} = (u_1, \ldots, u_n)$$

and

$$z_i = f_i(x_1, \ldots, x_n, \ y_1, \ldots, y_n), \quad u_i = g_i(x_1, \ldots, x_n), \quad i = 1, \ldots, n.$$

Since the group operations are continuous, so are the functions f_i and g_i. Hilbert's fifth problem is: Can ϕ be selected so that the functions f_i and g_i are not only continuous but also analytic? In other words: Is every locally Euclidean topological group G a Lie group?

The crucial step in proving that the answer is yes is due to Gleason [25], who showed that under a suitable condition on G every element x in G has a unique square root. These square roots are entirely analogous to our half-order iterates; proceeding as with the iteration, one can show that not only a unique $x^{1/2}$ but a unique x^r can be defined, for every real r. The construction of the one-parameter subgroups $\{x^r\}$ proved crucial in the eventual solution of Hilbert's fifth problem by Montgomery [51] and Zippin [52].

The condition on G which forces the uniqueness of square roots is that G should have no small subgroups: a neighborhood of the identity e should exist, which does not contain a subgroup of G other than $\{e\}$. This fitted in with the program of the attack on Hilbert's fifth problem since it was previously known that a Lie group cannot have small subgroups.

With its affirmative answer, Hilbert's fifth problem asserts, loosely speaking, that in a locally Euclidean topological group G certain minimal smoothness condition (namely, continuity) *together with the structure of G* implies certain maximal smoothness condition (namely, analyticity). A very much more elementary problem, in the same setting, occurs for functions of a complex variable: here it is shown that if such a function is differentiable then it is analytic.

We return now to our question of forcing a half-order iterate to be unique by imposing a property P. By using the Abel function $\phi(x)$ for $f(x)$, and adding to it a suitable $p(x)$, with $p(x) - x$ periodic of period 1, the reader may wish to show that taking P to mean $f \in C^\infty$ (i.e., f has continuous derivatives of all orders) is not sufficient. On the other hand, taking for P the property of being (real) analytic is too much—it may happen that no analytic half-order iterates $f_{1/2}$ exist for an analytic f.

By reference to Figure 3 the reader may wish to show that the double rectangle property illustrated there is a necessary and sufficient condition for f and g to commute: $f[g(x)] = g[f(x)]$. A construction analogous to the preceding one will show the following: given any order-preserving homeomorphism f of $[0, 1]$ onto itself, there exists a similar homeomorphism g which commutes with f and is essentially arbitrary on a subinterval.

As another connection with half-order iteration and its uniqueness we mention a generalization of Hilbert's seventh problem. This problem conjectured that if a and b are algebraic numbers, b is irrational, and $a \neq a^2$, then a^b is transcendental. This has been proved by Schneider and Gelfond [68]. The original question of Hilbert was: Is $2^{\sqrt{2}}$ transcendental? We may ask more generally: Is it true that the function $g(x) = x^{\sqrt{2}}$ assumes only transcendental values for algebraic x, if $x \neq x^2$? Observe the elementary proposition: if $x^2 = p$, where the positive integer p is not a perfect square, then x is irrational. By analogy, we formulate the following conjecture, which goes beyond Hilbert's problem: let $P(x)$ be a polynomial with integer coefficients, whose degree is not a perfect square, then there exists a unique "principal" half-order iterate $P_{1/2}(x)$ with the property of being transcendental for every algebraic x which is not a fixed point of P (i.e., not a root of $P(x) = x$). This

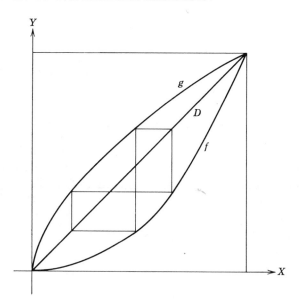

FIGURE 3. The double rectangle property.

conjecture may be generalized in various obvious and some unobvious ways.

Finally, the reader may wish to observe some well-known examples of the problem raised in the first sentence of this section. The factorial $n!$ is defined for nonnegative integers in the usual way. To extend factorials to real, and indeed, complex values of n we use the Γ-function, $\Gamma(x)$, defined by the following equivalent expressions:

$$\Gamma(x) = \int_0^\infty t^{x-1} e^{-t}\, dt \qquad \text{(the Euler integral, } \mathrm{Re}(x) > 0\text{),}$$

$$\frac{1}{\Gamma(x)} = xe^{\gamma x} \prod_{n=1}^\infty \left(1 + \frac{x}{n}\right) e^{-x/n} \qquad \begin{array}{l}\text{(the Weierstrass product, } x \text{ is an arbitrary} \\ \text{complex number),}\end{array}$$

$$\Gamma(x) = \lim_{n\to\infty} \frac{n!\, n^x}{x(x+1)\cdots(x+n)} \qquad \begin{array}{l}\text{(the Euler product, } x \text{ is a complex number} \\ \neq 0, -1, -2, \ldots\text{).}\end{array}$$

We then have $\Gamma(x+1) = x!$ if x happens to be a nonnegative integer. Here we also can formulate the uniqueness problem: What property of $\Gamma(x)$, in addition to some sufficient differentiability and the obvious recursion $\Gamma(x+1) = x\Gamma(x)$, will suffice to make it a unique extension of factorials? That some such property is necessary, we see by considering the example of an arbitrary sufficiently differentiable function $f(x)$ given for $0 \leq x \leq 1$, with $f(1) = 1$, and extending the domain of definition by putting

$$f(x+n) = (x+n-1)(x+n-2)\cdots(x+1)xf(x)$$

for $n = 1, 2, \ldots$; and then

$$f(y) = f(y + n)/[y(y + 1) \cdots (y + n)]$$

for $-n < y < 1 - n$. Now $f(x)$ is defined for all real x except at $x = 0, -1, -2, \ldots$ and yet it also satisfies the recursion $f(x + 1) = xf(x)$, the differentiability, and the normalization $f(1) = 1$, like $\Gamma(x)$.

As shown by Artin [3], the property of logarithmic convexity will do to make $\Gamma(x)$ a unique generalization of the factorials. Here a function $f(x)$ is called logarithmically convex if its logarithm is convex, i.e., if $ff'' - f'^2 \geq 0$. We observe that the concept of logarithmic convexity is related to that of order of growth: a positive logarithmically convex function increases faster than any exponential e^{ax}.

Once the factorial is uniquely extended to real or complex domain we may define derivatives of real or complex order a. Let $f(x) = x^n$, then $f'(x) = nx^{n-1}$, $f''(x) = n(n - 1)x^{n-2}$ and generally

$$f^{(k)}(x) = \frac{n!}{(n - k)!} x^{n-k}.$$

Therefore it is plausible to define for $g(x) = x^b$ the ath derivative by

$$g^{(a)}(x) = \frac{\Gamma(b + 1)}{\Gamma(b - a + 1)} x^{b-a}.$$

We may then extend this definition to a wider class of functions by the linearity property:

$$(c_1 g_1 + c_2 g_2)^{(a)} = c_1 g_1^{(a)} + c_2 g_2^{(a)}.$$

Alternatively, we may start with integration and consider the iterated indefinite integrals:

$$I_1(x) = \int_{x_0}^{x} f(x)\, dx, \qquad I_2(x) = \int_{x_0}^{x} I_1(x)\, dx, \qquad \ldots;$$

$$I_{n+1}(x) = \int_{x_0}^{x} I_n(x)\, dx, \qquad \ldots.$$

By induction or otherwise we find the explicit expression for the nth iterate:

$$I_n(x) = \frac{1}{n!} \int_{x_0}^{x} (x - y)^{n-1} f(y)\, dy$$

and, motivated by the above, we define

$$I_a(x) = \frac{1}{\Gamma(a + 1)} \int_{x_0}^{x} (x - y)^{a-1} f(y)\, dy.$$

Reversing the usual course by treating the derivative as the anti-integral, we may now define the ath order derivative by

$$D_a f(x) = \frac{1}{\Gamma(1-a)} \int_{x_0}^{x} (x-y)^{-a-1} f(y) \, dy.$$

5. NEWTON–RAPHSON METHOD AND ITERATION

Let us apply the Newton–Raphson method to finding the square root of a positive number A. We let $x_0 > 0$ be the initial approximation to A and we define

$$x_{n+1} = \frac{1}{2}\left(x_n + \frac{A}{x_n}\right), \qquad n = 0, 1, 2, \ldots$$

then the sequence $x_0, x_1, x_2, \ldots$ converges very rapidly to $\sqrt{A}$. The conjugate function method enables us not only to prove the convergence and estimate its rapidity but to obtain x_n explicitly. Let

$$f(x) = \frac{1}{2}\left(x + \frac{A}{x}\right), \qquad s(x) = x^2,$$

$$\phi(x) = \frac{x - \sqrt{A}}{x + \sqrt{A}}, \qquad \phi^{-1}(x) = \frac{1+x}{1-x}\sqrt{A}.$$

Then f and s are conjugate under $\phi \colon f = \phi^{-1} s \phi$, $f_n = \phi^{-1} s_n \phi$. Since $s_n(x) = x^{2^n}$ and $x_n = f_n(x_0)$, we have

$$x_n = \sqrt{A}\, \frac{(x_0 + \sqrt{A})^{2^n} + (x_0 - \sqrt{A})^{2^n}}{(x_0 + \sqrt{A})^{2^n} - (x_0 - \sqrt{A})^{2^n}}$$

and the rest follows. All this may be generalized slightly; instead of square roots we consider the roots of a quadratic equation. Let

$$q(x) = x^2 - (b+d)x + bd$$

and suppose that we solve $q(x) = 0$ by assuming an initial approximation x_0 and defining

$$x_{n+1} = x_n - \frac{q(x_n)}{q'(x_n)}, \qquad n = 0, 1, 2, \ldots.$$

We have then

$$x_{n+1} = f(x_n) \quad \text{where} \quad f(x) = \frac{x^2 - bd}{2x - b - d}.$$

Let $s(x) = x^2$ as before and let

$$\phi(x) = \frac{x-b}{x-d}, \qquad \phi^{-1}(x) = \frac{dx-b}{x-1};$$

then f and s are again conjugate under $\phi : f = \phi^{-1} s\phi$, and hence

$$x_n = \frac{d(x_0 - b)^{2^n} - b(x_0 - d)^{2^n}}{(x_0 - b)^{2^n} - (x_0 - d)^{2^n}}.$$

Without loss of generality let $d < b$. If $|x_0 - b| < |x_0 - d|$ then x_n converges to the larger root b; if $|x_0 - b| > |x_0 - d|$ then x_n converges to the smaller root d; if $|x_0 - b| = |x_0 - d|$ then there is no convergence; moreover the sequence $x_0, x_1, \ldots$ cannot even be continued beyond its initial term. In fact, it means that the tangent at the initial approximation x_0 is horizontal, hence parallel to the x-axis, as shown in Figure 1. We may, and even ought to, ask whether similar explicit expressions can be found in other cases. If we solve the equation $f(x) = 0$, by the Newton–Raphson method then the sequence of approximations is given by:

$$x_0, \ x_{n+1} = g(x_n), \ g(x) = x - f(x)/f'(x).$$

Hence, solving a simple differential equation,

$$f(x) = Ke^{\int dx/[x - g(x)]}. \tag{1}$$

It follows that if $g(x)$ is explicitly iterable then an explicit expression for the nth Newton–Raphson approximation x_n can be found for solving $f(x) = 0$ with $f(x)$ given by (1).

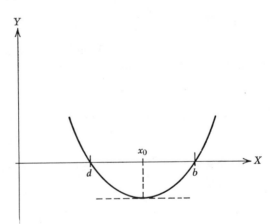

FIGURE 1.   Unfortunate initial guess.

6. APPROXIMATE ITERATION

It is the exception rather than the rule, to be able to find an explicit formula for the nth iterate $f_n(x)$ of a function $f(x)$. In this section we consider the case when $f(x) = \sin x$, $x_0 > 0$ is arbitrary number $\neq k\pi$, and we are interested in $x_n = f_n(x_0)$. Using the staircase diagram of Figure 1 we check that the sequence $x_0, x_1, \ldots$ is steadily decreasing and tends, rather slowly, to 0. The question is, how slowly? We derive an asymptotic formula due to de Bruijn [11], as follows. Using the power series for $\sin x$ we have

$$x_{n+1} = \sin x_n = x_n - x_n^3/6 + x_n^5/120 - \cdots \tag{1}$$

and squaring it

$$x_{n+1}^2 = x_n^2 - x_n^4/3 + 2x_n^6/45 - \cdots;$$

if we put $u_n = x_n^2/3$ then this becomes

$$u_{n+1} = u_n - u_n^2 + 2u_n^3/5 - \cdots$$

which is of the general form

$$u_{n+1} = u_n - u_n^2 + 0(u_n^3).$$

We also have $u_n \to 0$ for increasing n, and we let $v_n = 1/u_n$. It follows that $v_0, v_1, \ldots$ is an increasing sequence, it tends to $+\infty$, and

$$v_{n+1} - v_n = 1 + 0(1/v_n). \tag{2}$$

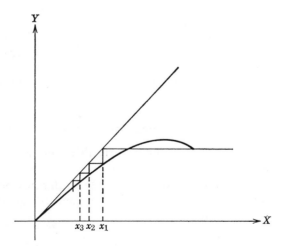

FIGURE 1.   Sine iteration.

This is very similar to the relation investigated at some length at the end of
the section on telescoping cancellation in the chapter on series and products,
and we shall use similar technique here. The reason for passage from x_n to
u_n to v_n was precisely to bring (1) into the form of (2) which is adapted to
telescoping cancellation. Writing 0, 1, $\ldots$, $n - 1$ in place of n in (2) and
adding, we have

$$v_n = n + 0\left(\sum_{k=0}^{n-1} 1/v_k\right)$$

from which it follows that

$$v_n = n + 0(\log n)$$

and hence

$$u_n = \frac{1}{n} + 0(n^{-2} \log n)$$

so that finally

$$x_n = \frac{\sqrt{3}}{n}\left[1 + 0\left(\frac{\log n}{n}\right)\right].$$

7. MULTIPLE RECURSIONS

Certain multiple recursions can be solved explicitly by similar methods. Suppose, for instance, that x_0 and y_0 are given and

$$x_{n+1} = x_n^2 - y_n^2, \qquad y_{n+1} = 2x_n y_n, \qquad n = 0, 1, 2, \ldots. \tag{1}$$

We put $z_n = x_n + iy_n$ so that z_0 is known and from (1) we have $z_{n+1} = z_n^2$ so
that $z_n = f_n(z_0)$. Therefore

$$x_n = \text{Re}(x_0 + iy_0)^{2^n}, \qquad y_n = \text{Im}(x_0 + iy_0)^{2^n}.$$

Similarly, if x_0 and y_0 are known and

$$x_{n+1} = ax_n^2 - ay_n^2 + 4x_n, \qquad y_{n+1} = 2ax_n y_n + 4y_n, \qquad n = 0, 1, 2, \ldots$$

we put $z_k = x_k + iy_k$ and we have

$$z_n = f_n(z_0) \quad \text{where} \quad f(z) = az^2 + 4z.$$

Since the nth iterate $f_n(z_0)$ can be found explicitly (as in one of the previous
examples) we can express x_n and y_n explicitly.

An interesting case of this type of iteration is the arithmetic-geometric mean
of Gauss. Let a_0 and b_0 be two distinct positive numbers and define the successive arithmetic and geometric means:

$$a_{n+1} = \frac{a_n + b_n}{2}, \qquad b_{n+1} = \sqrt{a_n b_n}, \qquad n = 0, 1, 2, \ldots. \tag{2}$$

It is clear that both sequences $a_0, a_1, a_2, \ldots$ and $b_0, b_1, b_2, \ldots$ converge to a common limit $d = d(a_0, b_0)$, for both sequences are bounded and monotone, and their limits must be equal. The value of d may be obtained, in a rather unobvious way, by using a special case of Landen's modular transformation on an elliptic integral. Let

$$K(a_1, b_1) = \int_0^{\pi/2} (a_1^2 \cos^2 \theta + b_1^2 \sin^2 \theta)^{-1/2} \, d\theta$$

be the complete elliptic integral of the first kind. Then

$$K(a_1, b_1) = \frac{1}{2a_1} \int_0^\pi (1 - k_1^2 \sin^2 \theta)^{-1/2} \, d\theta$$

where $k_1^2 = 1 - b_1^2/a_1^2$. We now introduce the new variable ϕ by

$$\tan \theta = \frac{\sin 2\phi}{k_1 + \cos 2\phi}$$

or, equivalently, by $k_1 \sin \theta = \sin(2\phi - \theta)$. As θ increases from 0 to π, ϕ increases from 0 to $\pi/2$ and so

$$K(a_1, b_1) = \frac{1}{a_1(1 + k_1)} \int_0^{\pi/2} (1 - k^2 \sin^2 \phi)^{-1/2} \, d\phi \tag{3}$$

where

$$k = \frac{2\sqrt{k_1}}{1 + k_1}.$$

Absorbing the constant outside the integral in (3) into the integrand and computing k, we find

$$K(a_1, b_1) = K(a_0, b_0),$$

where the a's and b's are as in (2). It follows that

$$K(a_0, b_0) = K(a_1, b_1) = K(a_2, b_2) = \cdots = K(d, d) = \frac{\pi}{2d}.$$

Hence the arithmetic-geometric mean $d(a_0, b_0)$ is given by the elliptic integral as

$$d(a_0, b_0) = \frac{\pi}{2K(a_0, b_0)}.$$

The convergence of the sequences defined by (2) is very rapid as shown in the following numerical example:

$$a_0 = 1.0000\ 0000\ 0000\ 0000 \qquad b_0 = 0.5000\ 0000\ 0000\ 0000$$
$$a_1 = 0.7500\ 0000\ 0000\ 0000 \qquad b_1 = 0.7071\ 0678\ 1186\ 5475$$
$$a_2 = 0.7285\ 5339\ 0593\ 2736 \qquad b_2 = 0.7282\ 3765\ 7560\ 9851$$
$$a_3 = 0.7283\ 9552\ 4077\ 1293 \qquad b_3 = 0.7283\ 9550\ 6969\ 7744$$
$$a_4 = 0.7283\ 9551\ 5523\ 4533 \qquad b_4 = 0.7283\ 9551\ 5523\ 4533.$$

For a variety of methods to compute elliptic integrals and elliptic functions by the arithmetic-geometric mean and related concepts, see reference [35].

The reader who can recognize a good thing when he sees one, may ask himself: Precisely what is behind this hyper-rapid convergence? Suppose that we replace (2) by the general relations

$$a_{n+1} = f(a_n, b_n), \qquad b_{n+1} = g(a_n, b_n) \tag{4}$$

where f and g satisfy the following conditions:

(a) both are homogeneous of degree 1,
(b) $f(x, x) = g(x, x) = x$,
(c) $f(1, z)$ and $g(1, z)$ are analytic near $z = 1$.

Let

$$f(1, 1 + z) = 1 + A_1 z + A_2 z^2 + \cdots$$
$$g(1, 1 + z) = 1 + B_1 z + B_2 z^2 + \cdots$$

then it can be proved that for suitable initial a_0 and b_0 the sequences defined by (4) have a common limit; moreover, the convergence gets more and more rapid the more equalities

$$A_1 = B_1, \qquad A_2 = B_2, \qquad \ldots$$

we have. In our example of the arithmetic-geometric mean

$$f(1, 1 + z) = 1 + \frac{1}{2} z, \qquad g(1, 1 + z) = (1 + z)^{1/2} = 1 + \frac{1}{2} z - \frac{1}{8} z^2 + \cdots$$

the equality of the first two coefficients is already sufficient for the speed shown in our numerical example. For further reference see [23] and [24].

Another example, somewhat similar to the arithmetic-geometric mean of Gauss, is the following. Let

$$I(a) = \int_0^{\pi/2} \log(1 + a \sin^2 \theta)\, d\theta, \qquad a > 0;$$

differentiating with respect to a, we find

$$aI'(a) = \frac{\pi}{2} - \int_0^{\pi/2} \frac{d\theta}{1 + a \sin^2 \theta} = \frac{\pi}{2}\left(1 - \frac{1}{\sqrt{a + 1}}\right)$$

so that

$$I(a) = \frac{\pi}{2} \log\left(a \frac{\sqrt{a+1}+1}{\sqrt{a+1}-1}\right) + C$$

and since

$$\lim_{a \to 0} a \frac{\sqrt{a+1}+1}{\sqrt{a+1}-1} = 4$$

and $I(0) = 0$, we find that

$$I(a) = \frac{\pi}{2} \log \frac{a(\sqrt{a+1}+1)}{4(\sqrt{a+1}-1)}. \tag{5}$$

Let

$$f(a) = \frac{a^2}{4(1+a)}; \tag{6}$$

re-writing (5) as

$$I(a) = \frac{\pi}{4} \log \frac{a^2(\sqrt{a+1}+1)^2}{16(\sqrt{a+1}-1)^2}$$

we verify that

$$I(a) - \frac{1}{2} I[f(a)] = \frac{\pi}{4} \log(1+a). \tag{7}$$

Replacing here a by $f(a)$ and dividing by 2, we have

$$\frac{1}{2} I[f(a)] - \frac{1}{4} I[f_2(a)] = \frac{\pi}{8} \log[1+f(a)]$$

which together with (7) gives us

$$I(a) - \frac{1}{4} I[f_2(a)] = \frac{\pi}{4} \log(1+a)[1+f(a)]^{1/2}.$$

Similarly,

$$I(a) - \frac{1}{8} I[f_3(a)] = \frac{\pi}{4} \log(1+a)[1+f(a)]^{1/2}[1+f_2(a)]^{1/4},$$

and generally

$$I(a) - 2^{-n} I[f_n(a)] = \frac{\pi}{4} \log \prod_{j=0}^{n-1} [1+f_j(a)]^{2^{-j}}. \tag{8}$$

We note for any $a > 0$ that $f_n(a) \to 0$ as $n \to \infty$. Passing to the limit on n in (8) we find that

$$I(a) = \frac{\pi}{4} \log \prod_{j=0}^{\infty} [1 + f_j(a)]^{2-j}$$

and comparing this with (5) we have

$$\prod_{j=0}^{\infty} [1 + f_j(a)]^{2-j} = e^{4I(a)/\pi} = \frac{a^2}{16} \frac{2 + a + 2\sqrt{1+a}}{2 + a - 2\sqrt{1+a}}. \tag{9}$$

It so happens that we can find an explicit expression for $f_n(a)$. For let $\phi(x) = 1/x$ so that also $\phi^{-1}(x) = 1/x$, then by (6)

$$\phi^{-1} f \phi = 4a(1 + a);$$

as shown before, the function $g(a) = 4a(1 + a)$ is explicitly iterable and

$$g_n(a) = \sinh^2 2^n \, \text{arg sinh} \, \sqrt{a}$$

so that

$$f_n(a) = \text{cosech}^2 2^n \, \text{arg cosech} \, \sqrt{a}$$

and (9) becomes

$$\prod_{n=0}^{\infty} [1 + \text{cosech}^2 2^n \, \text{arg cosech} \, \sqrt{a}]^{1/2^n} = \frac{a^2}{16} \frac{2 + a + 2\sqrt{1+a}}{2 + a - 2\sqrt{1+a}}.$$

Substituting here arg cosech $\sqrt{a} = t/2$ and expressing the hyperbolic functions by exponentials, and then letting $b = e^{-t}$, we have

$$\prod_{n=0}^{\infty} \left(\frac{1 + b^{2^n}}{1 - b^{2^n}} \right)^{1/2^n} = \frac{1}{(1-b)^2}, \qquad b < 1; \tag{10}$$

unlike (9) this is almost a triviality as can be seen by factorizing $1 - b^{2^n}$: we have then for the left-hand side of (10)

$$\frac{1+b}{1-b} \frac{(1+b^2)^{1/2}}{(1+b)^{1/2}(1-b)^{1/2}} \frac{(1+b)^{1/4}}{(1+b^2)^{1/4}(1+b)^{1/4}(1-b)^{1/4}} \cdots$$

which is $(1-b)^{-2}$ by checking the exponents of various factors.

8. AN APPLICATION TO PROBABILITY

Suppose that a single individual fissions at the time $t = 1$, splitting into 0, 1, 2, ... new individuals with respective probabilities $p_0, p_1, p_2, \ldots$. At the time $t = 2$ each of the new individuals undergoes the same process, and with the same probabilities, then the same happens at the times $t = 3, 4, \ldots$, and so on. Let p_{nk} be the probability of having exactly k individuals of the nth generation (i.e., at the time $n + 1/2$). We introduce the generating function

$$f(x) = p_0 + p_1 x + p_2 x^2 + \cdots, \qquad f(1) = 1,$$

and we find that the corresponding function for the nth generation probabilities is the nth iterate

$$f_n(x) = p_{n0} + p_{n1}x + p_{n2}x^2 + \cdots.$$

Suppose for instance that

$$f(x) = \frac{1}{1 + p - px}, \qquad p > 0,$$

so that the probability of fissioning into k individuals is $p^k/(1 + p)^{k+1}$. The nth iterate is easily found by induction:

$$f_n(x) = \frac{1 - p^n - (p - p^n)x}{1 - p^{n+1} - (p - p^{n+1})x}, \qquad p \neq 1,$$

$$f_n(x) = \frac{n - (n - 1)x}{n + 1 - nx}, \qquad p = 1.$$

Therefore

$$p_{nk} = \frac{p^n(1 - p)^2(p - p^{n+1})^{k-1}}{(1 - p^{n+1})^{k+1}}, \qquad p \neq 1,$$

$$p_{nk} = \frac{n^{k-1}}{(n + 1)^{k+1}}, \qquad p = 1.$$

Let N_n be the expected number of individuals in the nth generation:

$$N_n = \frac{df_n(x)}{dx}\bigg|_{x=1}.$$

Then, since $f_n(x) = f(f_{n-1}(x))$ we have by the chain-rule of differentiation

$$N_n = f'(1)N_{n-1}, \qquad N_n = [f'(1)]^n.$$

In the case we have considered $f'(1) = p$ and hence the expected number N_n is p^n. It follows that if $p > 1$, we have a population explosion; for $p < 1$, a certain extinction; and for $p = 1$, a stationary state.

9. ITERATION OF A FUNCTION OF SEVERAL VARIABLES

In every example so far the function f to be iterated was defined for, and took its values in, the same set: a subset of real or complex numbers, or a plane, etc. A problem, seemingly much harder, is how to iterate a function of several real variables, or more generally, a function whose domain and range are different sorts of spaces. We indicate some difficulties on a simple example.

Let A, B, C be three pairwise externally tangent circles in the plane whose curvatures, i.e., the reciprocals of the radii, are a, b, c. We inscribe a circle C_1, of curvature c_1, into the curvilinear triangle tangent to A, B, and C, as shown in Figure 1. The curvature c_1 is given by the Soddy formula

$$c_1 = a + b + c + 2(ab + ac + bc)^{1/2};$$

if the negative square root were taken here we should have obtained the curvature c_1' of the circle C_1' which is tangent to A, B, and C and contains them in its interior. We write the Soddy formula as

$$c_1 = F(a, b, c)$$

and we inscribe three further circles C_2, C_3, C_4 tangentially to their predecessors as shown in Figure 1. Their curvatures are then

$$c_2 = F(F(a, b, c), b, c), \qquad c_3 = F(a, F(a, b, c), c), \qquad c_4 = F(a, b, F(a, b, c));$$

the expressions on the right-hand sides may be called the three partial second-order iterates of $F(a, b, c)$. Inscribing similarly nine further circles $C_5, \ldots, C_{13}$, of curvatures $c_5, \ldots, c_{13}$, tangentially into the interstices as shown in the figure, we find for their curvatures:

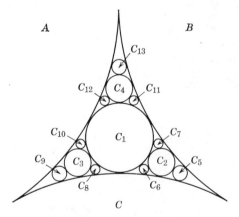

FIGURE 1.   Tangent circles.

$$c_5 = F(F(F(a, b, c), b, c), b, c);$$
$$c_6 = F(F(a, b, c), F(F(a, b, c), b, c), c);$$
$$c_7 = F(F(a, b, c), b, F(F(a, b, c), b, c));$$
$$c_8 = F(F(a, F(a, b, c), c), F(a, b, c), c);$$
$$c_9 = F(a, F(a, F(a, b, c), c), c);$$
$$c_{10} = F(a, F(a, b, c), F(a, F(a, b, c), c));$$
$$c_{11} = F(F(a, b, F(a, b, c)), b, F(a, b, c));$$
$$c_{12} = F(a, F(a, b, F(a, b, c)), F(a, b, c));$$
$$c_{13} = F(a, b, F(a, b, F(a, b, c))).$$

The right-hand sides are the nine partial third-order iterates of $F(a, b, c)$. This inscription and iteration process may be continued in the obvious fashion. We notice that in iterating a function $f(x)$ of one variable it sufficed to use the integers 0, 1, 2, 3, ... for the integral order iteration and the reals for real-order iteration; both structures have *simple linear order*. On the other hand, in our example it is not even clear what sort of order-structure should be used to label the partial iterates. In particular, it is not at all obvious how to define partial iterations of fractional order. However, we can compute the successive curvatures $c_1, c_2, c_3, \ldots$ of the circles shown in Figure 2. For we have now by the Soddy formula

$$c_0 = c, \quad c_{n+1} = a + b + c_n + 2(ab + ac_n + bc_n)^{1/2} = f(c_n)$$

say, and so we have to iterate a function of only one variable: $c_n = f_n(c)$. We apply the method of conjugate functions and find the Abel function for $f(x)$:

$$f(x) = h[\psi(h^{-1}(x))]$$

where $\psi(x) = x + 1$ and $h(x) = (a + b)x^2 - ab/(a + b)$. Hence

$$c_n = (a + b)n^2 + 2n[(a + b)c + ab]^{1/2} + c.$$

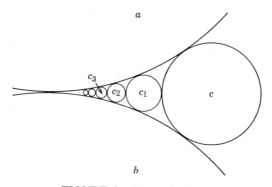

FIGURE 2.   Tangent circles.

10. ITERATION AND ORDERS OF MAGNITUDE

We may regard multiplication as iterated addition: let $f(x) = x + a$, then $f_b(0) = ab$; call the latter function $f_1(a, b)$. Next, exponentiation is iterated multiplication: if $g(x) = f_1(a, x)$, then $g_b(1) = a^b$; we put $f_2(a, b) = a^b$. Similarly, to continue the process we put $h(x) = f_2(a, x)$ and we define $f_3(a, b)$ by $f_3(a, b) = h_b(1)$ so that

$$f_3(a, b) = a^{a^{\cdot^{\cdot^{\cdot^a}}}} \qquad (b \text{ occurrences of } a).$$

Continuing in this fashion, we obtain the Ackermann sequence

$$f_1(a, b), f_2(a, b), f_3(a, b), \ldots$$

of functions, each of which increases faster than any bounded iterate of the previous one. To obtain $f_{n+1}(a, b)$ from $f_n(a, b)$ we use:

$$k(x) = f_n(a, x), \qquad f_{n+1}(a, b) = k_b(1).$$

Diagonalizing the Ackermann sequence by the Cantor diagonal process yields the Ackermann function:

$$A(n) = f_n(n, n).$$

We have then $A(1) = 1$, $A(2) = 2^2$, $A(3) = 3^{3^3}$, and

$$A(4) = 4^{4^{\cdot^{\cdot^{\cdot^4}}}}$$

(the number of occurrences of 4 is $4^{4^{\cdot^{\cdot^{\cdot^4}}}}$ (the number of occurrences of 4 is $4^{4^{4^4}}$)). There is another, though related, way of similarly generating fast increasing functions. It has a more analytical emphasis and is concerned with real rather than with integer variables. We observe that from the power series

$$e^x = \sum_{n=0}^{\infty} x^n/n!$$

it follows that e^x increases faster than any power x^k (or indeed faster than any finite iterate of x^k) since the series for e^x contains terms like $x^{k+1}/(k + 1)!$ (or like $x^{k^m+1}/(k^m + 1)!$). With this as a guide we start with some moderately fast-growing function $f(x)$ (x^2 will do, or e^x) and we define

$$g(x) = \sum_{n=1}^{\infty} f_n(x)/f_n(n) \qquad (1)$$

where f_n denotes, as usual, the nth iterate. From (1) it follows that $g(x)$ increases faster than any iterate of $f(x)$; now we continue the process:

$$h(x) = \sum_{n=1}^{\infty} g_n(x)/g_n(n), \tag{2}$$

$$k(x) = \sum_{n=1}^{\infty} h_n(x)/h_n(n), \tag{3}$$

and so on. All the series will converge and we obtain the sequence $f(x)$, $g(x)$, $h(x)$, $k(x)$, ... which we re-label as $F_1(x)$, $F_2(x)$, $F_3(x)$, ... Applying the diagonal process, we let

$$F(x) = \sum_{n=1}^{\infty} F_n(x)/F_n(n)$$

and we obtain a real-valued analogue of the Ackermann function $A(n)$. The reader may wish to compare the growth of the two functions, starting $F(x)$ with, say, $f(x) = x^2$.

It may occur to the reader to ask: Why bother with such rather unnatural-seeming constructs as the Ackermann function $A(n)$? Besides the natural fascination with bigness, there is another reason for $A(n)$. Let us limit ourselves to functions, of any finite number of variables, whose domain and range are natural numbers (or n-tuples of such). Let us define the primitive recursive functions to be the smallest system S of functions, which contains the "initial" functions

$$f(x) = x + 1 \qquad \text{(Peano successor)}$$
$$f(x_1, \ldots, x_n) = k \qquad \text{(constants)} \tag{4}$$
$$f(x_1, \ldots, x_n) = x_i \qquad \text{(projections)}$$

and which is closed under substitution:

$$\text{if } F, f_1, \ldots, f_n \in S \text{ then } g = F[f_1(x_1, x_2, \ldots), \ldots, f_n(x_1, x_2, \ldots)] \in S, \tag{5}$$

and under primitive recursion:

$$\text{if } f(x_2, \ldots, x_n) \qquad \text{and} \qquad g(x, x_1, \ldots, x_n) \tag{6}$$

are in S then the function $F(x_1, \ldots, x_n)$, defined by

$$F(0, x_2, \ldots, x_n) = f(x_2, \ldots, x_n),$$
$$F(x + 1, x_2, \ldots, x_n) = g(x, F(x, x_2, \ldots, x_n), x_2, \ldots, x_n),$$

is also in S. The system S contains many functions ordinarily met with and whatever formalization of "computability" we use, we should like our computable functions to include the primitive recursive functions S. The natural question arises now whether there exist functions which qualify to be computable, but which are not primitive recursive. As it turns out, no primitive recursive function can increase as fast as $A(n)$, so that $A(n)$ is computable but not primitive recursive. That was precisely the reason why Ackermann has introduced $A(n)$. However, it is easy to give examples of computable functions which are neither primitive recursive nor fast-growing. In fact, such a function $f(n)$ may take up two values 0, 1 only: $f(n) = 0$ if an integer k exists such that $A(k) = n, f(n) = 1$ otherwise.

Finally, we observe that in order to obtain the class R of recursive, or computable, functions, it is necessary to add just one further requirement to (4), (5), and (6): R is the smallest system containing the initial functions (4), and closed under substitution (5), primitive recursion (6), and one further schema, the minimalization [36]. This one is as follows:

$$\text{if } f(x_1, \ldots, x_n, y) \text{ is any function in } R, \tag{7}$$

such that for every set of values $x_1, \ldots, x_n$ a value of y exists for which $f(x_1, \ldots, x_n, y) = 0$, then the function

$$g(x_1, \ldots, x_n) = \min\{y : f(x_1, \ldots, y_n, y) = 0\}$$

is also in R.

11. ITERATION AND RELIABLE CIRCUITS

An electromagnetic relay, illustrated in Figure 1, is a device with a coil; when the coil is energized the contact at C is made and the circuit AB is closed. It may happen that the relay is faulty; it produces contact with unenergized coil, and it leaves open contact with energized coil. Assume the following description of a faulty relay which gives the probabilities of the four principal events:

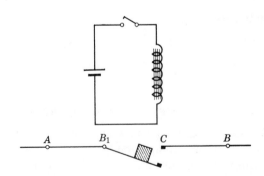

FIGURE 1.   Schematic relay.

if coil energized 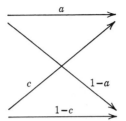 then contact closed

if coil unenergized then contact open

We may suppose that $a \geq c$, for otherwise we reverse the roles of open-closed and energized-unenergized.

Suppose now that there is a large number of identically faulty relays with the above probabilities. Let their coils be energized or unenergized simultaneously, and suppose that the relays function (or fail to) in a statistically independent fashion. Under these conditions Moore and Shannon proved that if $a > c$ then it is possible to synthesize out of sufficiently many identically faulty relays a network circuit which behaves like a perfect relay with probability arbitrarily close to 1.

It is necessary to examine the meaning of the last phrase. We consider a two-terminal network with the terminals A and B, made up of our relays each of which has, independently, the probability p of being closed. Let the reliability function $h(p)$ of the network be the probability that there is a contact between A and B. Now the Moore-Shannon design is as follows: let $a > c$ and let $\delta > 0$ be arbitrarily small, then there is a network whose realiability function $h(p)$ has the graph shown in Figure 2. That is, we have

$$0 \leq h(p) \leq \delta \quad \text{for} \quad 0 \leq p \leq c, \qquad 1 - \delta \leq h(p) < 1 \quad \text{for} \quad a \leq p \leq 1.$$

In words: when the coils are unenergized the probability of contact is arbitrarily small, when the coils are energized the probability of contact is arbitrarily close to 1. Briefly: the network acts like a perfect relay.

Let N_1 and N_2 be two networks with reliability functions $h_1(p)$ and $h_2(p)$. We can place them in series which gives us a network with reliability function $h_1(p)h_2(p)$; we can place them in parallel which results in a new network with reliability function

$$h_1(p) + h_2(p) - h_1(p)h_2(p) = 1 - [1 - h_1(p)][1 - h_2(p)].$$

We can also iterate N_2 over N_1 by replacing each relay of N_1 with a copy of N_2: the new network has reliability function $h_1(h_2(p))$. In particular, if a network with the function $h(p)$ is iterated over itself n times the new network has reliability function $h_n(p)$ (the nth iterate of $h(p)$).

Moore and Shannon prove first that except for the special case of a single relay-network (with $h(p) = p$) each reliability function $h(p)$ satisfies the following conditions:

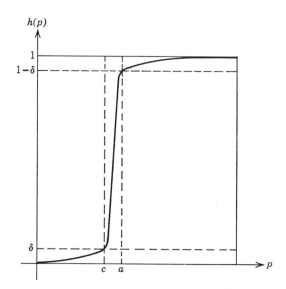

FIGURE 2.   Reliability function of a reliable network.

(a) $h(0) = 0$, $h(1) = 1$, $h(p)$ strictly increasing for $0 \le p \le 1$,

(b) $h(p) = p$ has at most one root p_0 with $0 < p_0 < 1$,

(c) If such fixed point p_0 exists then $h(p) < p$ for $0 < p < p_0$ and $h(p) > p$ for $p_0 < p < 1$.

By a preliminary explicit construction, using series, parallel, and some other compositions, they then show that there is a network N whose internal fixed point p_0 satisfies $c < p_0 < a$. If this N is iterated over itself n times and n is sufficiently high, then there results the arbitrarily reliable network of Figure 2. For we merely observe that if $h(p)$ is subdiagonal ($h(p) < p$ for $0 < p < p_0$) .then

$$\lim_{n \to \infty} h_n(p) = 0, \qquad 0 \le p < p_0$$

(though not uniformly in p), while if $h(p)$ is superdiagonal (that is, $h(p) > p$ for $p_0 < p < 1$) then

$$\lim_{n \to \infty} h_n(p) = 1, \qquad p_0 < p \le 1.$$

3

SERIES AND PRODUCTS

1. TELESCOPING CANCELLATION

The formulas

$$(a_2 - a_1) + (a_3 - a_2) + \cdots + (a_n - a_{n-1}) = a_n - a_1$$
$$(a_2/a_1)(a_3/a_2) \cdots (a_n/a_{n-1}) = a_n/a_1$$

can be usefully applied in a variety of problems. A very simple example is in evaluating the sum

$$S = \sum_{k=1}^{\infty} \frac{1}{k(k+1)};$$

the nth partial sum is

$$s_n = \sum_{k=1}^{n} \frac{1}{k(k+1)} = \sum_{k=1}^{n} \left(\frac{1}{k} - \frac{1}{k+1} \right)$$

which by cancellation gives us $s_n = 1 - [1/(n+1)]$. Hence $S = 1$. The corresponding multiplicative example occurs in evaluating the product

$$P = \prod_{k=2}^{\infty} \left(1 - \frac{1}{k^2} \right) = \prod_{k=2}^{\infty} \frac{(k-1)(k+1)}{k^2} = \frac{1}{2}.$$

Observing that

$$\text{arc tan } f(n) - \text{arc tan } f(n+1) = \text{arc tan } \frac{f(n) - f(n+1)}{1 + f(n)f(n+1)}$$

we find by the telescoping cancellation that

$$\sum_{n=1}^{\infty} \text{arc tan } \frac{f(n) - f(n+1)}{1 + f(n)f(n+1)} = \text{arc tan } f(1)$$

provided that the series converges. For instance, with $f(n) = 1/n$ we have

$$\sum_{n=1}^{\infty} \arctan \frac{1}{1 + n + n^2} = \frac{\pi}{4}$$

and with $f(n) = 1/(2n - 1)$

$$\sum_{n=1}^{\infty} \arctan \frac{1}{2n^2} = \frac{\pi}{4}.$$

To sum the geometric progression

$$s = a + ar + ar^2 + \cdots + ar^n$$

we write

$$s = a + ar + ar^2 + \cdots + ar^n$$
$$rs = \quad ar + ar^2 + \cdots + ar^n + ar^{n+1}$$

so that subtracting and canceling,

$$s = a \frac{r^{n+1} - 1}{r - 1}.$$

Sums of powers

$$S_k(n) = \sum_{j=1}^{n} j^k$$

can be similarly evaluated. For instance, starting with the identity $(j + 1)^2 - j^2 = 2j + 1$, we have

$$(n + 1)^2 - n^2 = 2n + 1$$
$$n^2 - (n - 1)^2 = 2(n - 1) + 1$$
$$\vdots$$
$$2^2 - 1^2 = 2 \cdot 1 + 1,$$

adding and canceling we have

$$(n + 1)^2 - 1^2 = 2S_1(n) + n$$

whence $S_1(n) = n(n + 1)/2$. Suppose that $S_1(n), S_2(n), \ldots, S_{k-1}(n)$ are known; to evaluate $S_k(n)$ we use the binomial theorem with the exponent $k + 1$:

$$(n + 1)^{k+1} - n^{k+1} = \sum_{i=1}^{k+1} \binom{k + 1}{i} n^{k+1-i}$$

$$n^{k+1} - (n - 1)^{k+1} = \sum_{i=1}^{k+1} \binom{k + 1}{i} (n - 1)^{k+1-i}$$

$$\vdots$$

$$2^{k+1} - 1^{k+1} = \sum_{i=1}^{k+1} \binom{k + 1}{i} 1^{k+1-i},$$

adding and canceling we have

$$(k + 1)S_k(n) = (n + 1)^{k+1} - \sum_{i=2}^{k+1} \binom{k + 1}{i} S_{k+1-i}(n) - 1.$$

By approximating to a given series with a suitable telescoping series we may improve convergence. Suppose for instance that we wish to evaluate

$$S = \sum_{n=1}^{\infty} \frac{1}{n^3};$$

we have

$$S_1 = \sum_{n=1}^{\infty} \frac{1}{n(n + 1)(n + 2)} = \sum_{n=1}^{\infty} \left(\frac{1/2}{n} - \frac{1}{n + 1} + \frac{1/2}{n + 2} \right)$$

which may be written as

$$S_1 = \frac{1/2}{1} - \frac{1}{2} + \frac{1/2}{3} + \frac{1/2}{2} - \frac{1}{3} + \frac{1/2}{4} + \frac{1/2}{3} - \frac{1}{4} + \frac{1/2}{5} + \cdots$$

so that after cancellations

$$S_1 = \frac{1/2}{1} - \frac{1/2}{2} = \frac{1}{4}.$$

Here instead of the cancellation formula

$$(a_2 - a_1) + (a_3 - a_2) + \cdots + (a_n - a_{n-1}) = a_n - a_1$$

which uses first-order differences we use

$$(a_3 - 2a_2 + a_1) + (a_4 - 2a_3 + a_2) + \cdots$$
$$+ (a_n - 2a_{n-1} + a_{n-2}) = a_n - a_{n-1} - a_2 + a_1$$

which similarly uses second-order differences. Subtracting our two series we have

$$S - S_1 = S - \frac{1}{4} = \sum_{n=1}^{\infty} \left[\frac{1}{n^3} - \frac{1}{n(n + 1)(n + 2)} \right]$$

or

$$S = \frac{1}{4} + \sum_{n=1}^{\infty} \frac{3n + 2}{n^3(n + 1)(n + 2)} = \frac{1}{4} + 3 \sum_{n=1}^{\infty} \frac{1}{n^3(n + 2)}$$

$$- \sum_{n=1}^{\infty} \frac{1}{n^3(n + 1)(n + 2)}.$$

Since the terms of the last two series diminish faster than those of the original series, the above form of S is better adapted for numerical computation. We can, of course, apply similar procedure to each of the last two series to express S in a still faster converging form. However, we do better by using the centered form of telescoping cancellation:

$$S = 1 + \sum_{n=2}^{\infty} \frac{1}{n^3}, \qquad \sum_{n=2}^{\infty} \frac{1}{(n-1)n(n+1)} = \sum_{n=1}^{\infty} \frac{1}{n(n+1)(n+2)} = \frac{1}{4};$$

therefore, subtracting,

$$S = \frac{5}{4} - \sum_{n=2}^{\infty} \left(\frac{1}{n^3 - n} - \frac{1}{n^3} \right) = \frac{5}{4} - \sum_{n=2}^{\infty} \frac{1}{n^3(n^2 - 1)}.$$

Here the nth term is approximately n^{-5} whereas before it was $3n^{-4}$.

The following transformation of series, due to Kummer, uses a similar idea; suppose that:

(a) $S = \sum_{n=0}^{\infty} a_n$ is convergent with $a_n \neq 0$,

(b) $C = \sum_{n=0}^{\infty} c_n$ is convergent with known sum C,

(c) $\lim_{n \to 0} a_n/c_n = b$ exists and is $\neq 0$, then

$$S = bC + \sum_{n=0}^{\infty} (1 - bc_n/a_n)a_n \tag{1}$$

which transforms S into another series converging faster (because $1 - bc_n/a_n \to 0$). This may be repeated again on the right-hand side of (1). In this way we get

$$\sum_{n=1}^{\infty} \frac{1}{n^2} = 1 + \frac{1}{2^2} + \cdots + \frac{1}{k^2} + k! \sum_{n=1}^{\infty} \frac{1}{n^2(n+1)(n+2)\cdots(n+k)},$$

$$\sum_{n=0}^{\infty} \frac{1}{(n+a)^2(n+a+1)^2 \cdots (n+a+k-1)^2}$$

$$= \frac{2a + 3k - 2}{(4k-2)a^2(a+1)^2 \cdots (a+k-1)^2}$$

$$+ \frac{k^3}{4k-2} \sum_{n=0}^{\infty} \frac{1}{(n+a)^2(n+a+1)^2 \cdots (n+a+k)^2},$$

$$a \text{ real and } \neq 0, -1, -2, \ldots.$$

In the transformation due to Markov a convergent series $S = a_0 + a_1 + \cdots$ is transformed by representing the nth term itself as a convergent series:

$$S = a_{00} + a_{01} + a_{02} + \cdots \qquad \text{(summing up to } a_0)$$
$$+ a_{10} + a_{11} + a_{12} + \cdots \qquad \text{(summing up to } a_1)$$
$$+ \cdots$$

and now summing the above double array vertically rather than horizontally:

$$S = (a_{00} + a_{10} + a_{20} + \cdots) + (a_{01} + a_{11} + \cdots) + \cdots.$$

For instance, starting with

$$S = \sum_{n=1}^{\infty} \frac{1}{n^2}$$

we write

$$\frac{1}{k^2} = \frac{0!}{k(k+1)} + \frac{1!}{k(k+1)(k+2)} + \frac{2!}{k(k+1)(k+2)(k+3)} + \cdots$$

and we sum vertically. Using telescoping cancellation we evaluate the vertical sums obtaining

$$\sum_{n=1}^{\infty} \frac{1}{n^2} = 3 \sum_{n=1}^{\infty} \frac{(n-1)!(n-1)!}{(2n)!} \tag{2}$$

which transforms $\sum_{1}^{\infty} 1/n^2$ into a much faster convergent series. We observe here that (2) is a special case of the expansion

$$\left(\arcsin \frac{x}{2}\right)^2 = \frac{1}{2} \sum_{n=1}^{\infty} \frac{(n-1)!(n-1)!}{(2n)!} x^{2n}.$$

This is derived by integrating the series for $(1 - y^2)^{-1/2} \arcsin y$, given in Section 9.

A rather more spectacular result is obtained by starting similarly with $\sum_{n=1}^{\infty} 1/n^3$; we get eventually

$$\sum_{n=1}^{\infty} \frac{1}{n^3} = \sum_{n=1}^{\infty} (-1)^{n+1} \frac{(n-1)!(n-1)!}{(3n-2)!} \left[\frac{1}{(2n-1)^2} + \frac{5}{12n(3n-1)} \right] \tag{3}$$

where thirteen terms will give a twenty-digit accuracy. We observe here that the left-hand side of (3) is $\zeta(3)$ and that it is still unknown whether $\zeta(3)$ is irrational or not.

The simple additive form of telescoping cancellation could be stated thus: if

(a) $S = \sum_{n=0}^{\infty} a_n$ is a suitably convergent series, and

(b) $a_n = b_n - b_{n+1}$ $(n = 0, 1, 2, \ldots)$, then $S = b_0$.

A generalization is obtained replacing (b) by

(b₁) $a_n = b_n - b_{n+k}$ $(n = 0, 1, 2, \ldots)$, k is a positive integer, and $\lim b_n = B$. We have then

$$S_m = \sum_0^m a_j = \sum_{i=0}^{k-1} b_i - \sum_{i=m-k+1}^m b_{k+i}$$

so that passing to the limit on m we have

$$\sum_{n=0}^\infty a_n = b_0 + b_1 + \cdots + b_{k-1} - kB.$$

Still further generalization is obtained when we replace the condition (b) by
(b₂) $k \geq 2$ is a positive integer and

$$a_n = \sum_{i=1}^k c_i b_{n+i}, \qquad n = 0, 1, \ldots,$$

where

$$\sum_1^k c_i = 0 \qquad \text{and} \qquad \lim_{n \to \infty} b_n = B.$$

We have now

$$\sum_{n=0}^\infty a_n = \sum_{i=1}^k \left(\sum_{j=1}^i c_j \right) b_i + B \sum_{i=2}^k (i-1)c_i.$$

The digamma function $\psi(x)$ is defined as

$$\psi(x) = \sum_{n=1}^\infty \frac{x}{n(n+x)} - \gamma$$

where $\gamma = 0.577215665 \ldots$ is Euler's constant; equivalently, we have

$$\psi(x) = \frac{\Gamma'(x+1)}{\Gamma(x+1)}$$

where $\Gamma(x)$ is the Γ-function (occasionally the function $\Gamma'(x)/\Gamma(x)$ is called the digamma function; the name "digamma" derives from an archaic letter of the Greek alphabet, looking like capital F). Since

$$\frac{x}{n(n+x)} = \frac{1}{n} - \frac{1}{n+x}$$

we have by the telescoping cancellation

$$\psi(x+1) - \psi(x) = \sum_{n=1}^\infty \left(\frac{1}{n+x} - \frac{1}{n+x+1} \right) = \frac{1}{x+1}. \qquad (4)$$

Differentiating k times, we have

$$\psi^{(k)}(x) = (-1)^{k+1} k! \sum_{n=1}^{\infty} \frac{1}{(n+x)^{k+1}}, \qquad k = 1, 2, \ldots,$$

hence

$$\psi^{(k)}(x+1) - \psi^{(k)}(x) = (-1)^k \frac{k!}{(1+x)^{k+1}}.$$

Digamma function and its derivatives enable us to express in closed form sums of convergent series of the type

$$\sum_{n=1}^{\infty} \frac{P(n)}{Q(n)}$$

where P and Q are polynomials. For instance, suppose that we wish to evaluate

$$S = \sum_{n=1}^{\infty} \frac{1}{(n+a)(n+b)};$$

the nth term u_n has the partial fraction representation

$$u_n = \frac{1}{b-a} \left(\frac{1}{n+a} - \frac{1}{n+b} \right)$$

which may be written, using (4), as

$$u_n = \frac{1}{b-a} [\psi(n+a) - \psi(n+a-1) + \psi(n+b-1) - \psi(n+b)].$$

Now the nth partial sum $s_n = \sum_1^n u_j$ is obtained from the above by telescoping cancellation:

$$s_n = \frac{1}{b-a} [\psi(b) - \psi(a) + \psi(n+a) - \psi(n+b)].$$

By the series definition of $\psi(x)$ we have

$$\psi(n+a) - \psi(n+b) = (a-b) \sum_{k=1}^{\infty} \frac{1}{(n+k+a)(n+k+b)}$$

which tends to 0 as n increases. Hence

$$S = \frac{\psi(b) - \psi(a)}{b-a}.$$

By passage to the limit $(b \to a)$ we find

$$\sum_{n=1}^{\infty} \frac{1}{(n+a)^2} = \psi'(a).$$

The digamma function occurs in connection with the generalized Goldbach–Euler series:

$$\sum_{n=2}^{\infty} \sum_{s=0}^{\infty} \frac{1}{(ps+r)^n - 1} = \frac{1}{p}\left[\psi\left(\frac{r}{p}\right) - \psi\left(\frac{r-1}{p}\right)\right] + \frac{1}{r(r-1)} \tag{5}$$

where p is a positive integer and $r = p$ or $r = p + 1$; we exclude the case $r = p = 1$. Starting with the formula

$$\frac{1}{a(a-1)} = \sum_{n=2}^{\infty} a^{-n}$$

valid for $a > 1$ we form the sum

$$\sum_{a} \frac{1}{a(a-1)} = \sum_{a} \sum_{n=2}^{\infty} a^{-n} \tag{6}$$

where the summation $\sum_a$ extends over all numbers of the form $a = ps + r$, $s = 0, 1, 2, \ldots$, but it excludes all the numbers of the form $a = (pj + r)^n$ with $n \geq 2$ and $j = 0, 1, 2, \ldots$. Under the conditions on p and r the double sum in (6) will include every term of the form $(ps + r)^{-n}$ exactly *once* so that

$$\sum_{s=0}^{\infty} \frac{1}{(ps+r)(ps+r-1)}$$

$$- \sum_{n=2}^{\infty} \sum_{s=0}^{\infty} \frac{1}{(ps+r)^n[(ps+r)^n - 1]} = \sum_{n=2}^{\infty} \sum_{s=0}^{\infty} \frac{1}{(ps+r)^n}.$$

Putting together the two double series and observing that

$$\frac{1}{a(a-1)} + \frac{1}{a} = \frac{1}{a-1}$$

we find

$$\sum_{n=2}^{\infty} \sum_{s=0}^{\infty} \frac{1}{(ps+r)^n - 1} = \sum_{s=0}^{\infty} \frac{1}{(ps+r)(ps+r-1)}.$$

Breaking up the terms of the series on the right by partial fractions and using the digamma function we obtain (5). As special cases we have for $r = 2, p = 1$ the Goldbach series

$$\sum_{n=2}^{\infty} \sum_{s=0}^{\infty} \frac{1}{(s+2)^n - 1} = 1,$$

for $r = p = 2$ the Euler series

$$\sum_{n=2}^{\infty} \sum_{s=0}^{\infty} \frac{1}{(2s + 2)^n - 1} = \log 2,$$

and for $p = r = 3$

$$\sum_{n=2}^{\infty} \sum_{s=0}^{\infty} \frac{1}{(3s + 3)^n - 1} = \frac{1}{3} \left(\frac{3}{2} \log 3 - \frac{\pi}{2 \cdot 3^{1/2}} \right).$$

(*Note:* this is the same Goldbach who conjectured in a letter to Euler that every even number is a sum of two primes.)

   The principal idea in this section is that to sum $S = a_1 + a_2 + \cdots + a_n$ we represent a_j as the difference $a_j = b_{j+1} - b_j$, and we have then $S = b_{n+1} - b_1$. The situation is entirely analogous to that in calculus, where a definite integral

$$I = \int_a^b f(x) \, dx$$

is defined as the limit of certain sums; however, to evaluate I we do not use that definition but we employ the primitive or anti-derivative $F(x)$, given by $F'(x) = f(x)$, and we have then $I = F(b) - F(a)$. To push this analogy further the reader may observe that the digamma function plays in summation and differencing the same role that the function $\log x$ does in integration and differentiation. One may view the telescoping cancellation as a very elementary application of the motto " Man muss immer umkehren" ("One must always invert") of C. G. J. Jacobi. Jacobi was primarily interested in obtaining elliptic functions by inverting elliptic integrals; an elementary example of such inversion is that of the trigonometric function $\sin x$ as the function inverse to

$$\arcsin x = \int_0^x \frac{dt}{\sqrt{1 - t^2}}.$$

Here we are interested in *summation* of a_n and we do it by applying the *inverse* operation of *differencing*: $a_n = b_{n+1} - b_n$.

   An example of a somewhat different use of cancellation is found in generating rationally an approximation to the square-root function. Let $x_1 = 1$ and put

$$x_{n+1} = x_n + 1/x_n \qquad n = 1, 2, \ldots; \tag{7}$$

by writing this for $n = 1, 2, \ldots$, adding and canceling, we have

$$x_{n+1} = x_1 + \sum_{i=1}^{n} \frac{1}{x_i}. \tag{8}$$

Squaring the basic recurrence (7) and repeating the same steps we find

$$x_{n+1}^2 = x_1{}^2 + 2n + \sum_{i=1}^n \frac{1}{x_i{}^2}.$$

Therefore, $x_{n+1}^2 \geq 2n + 2$ and so $x_n \geq \sqrt{2n}$ for $n \geq 2$. Hence also $1/x_n \leq 1/\sqrt{2n}$ and therefore by (8)

$$x_{n+1} \leq \sum_{i=1}^n \frac{1}{\sqrt{2i}} + 2 - 2^{-1/2}. \tag{9}$$

Now we use the argument of the comparison of a sum and an integral (as in the integral test) and find

$$\sum_{i=1}^n \frac{1}{\sqrt{2i}} \leq 2^{-1/2} - 2^{1/2} + \varepsilon_n + \sqrt{2(n+1)}$$

where $\varepsilon_n \to 0$ as $n \to \infty$. Hence by (9)

$$x_n \leq \sqrt{2n} + 0.5858 \qquad n \text{ large.}$$

Putting together our estimates we have for large n

$$\sqrt{2n} \leq x_n \leq \sqrt{2n} + 0.5858.$$

Actually, more can be proved. Since for $n \geq 2$ we have $\sqrt{2n} \leq x_n$,

$$x_{n+1}^2 = 1 + 2n + \sum_{i=1}^n x_i{}^{-2} \leq \frac{3}{2} + 2n + \sum_{i=1}^n (2i)^{-1},$$

so that

$$x_{n+1}^2 \leq 2n + \frac{1}{2} \log n + K$$

for some constant K. Therefore also

$$x_n{}^2 \leq 2n + \frac{1}{2} \log n + K_1$$

and hence

$$\sqrt{2n} \leq x_n \leq \sqrt{2n + \frac{1}{2} \log n + K_1}.$$

Subtracting $\sqrt{2n}$ we have

$$0 \leq x_n - \sqrt{2n} \leq \sqrt{2n + \frac{1}{2} \log n + K_1} - \sqrt{2n}$$

and the last expression tends to 0 as n increases. Therefore

$$x_n - \sqrt{2n} \to 0 \quad \text{as} \quad n \to \infty.$$

Our procedure here may be schematized as follows. Let $m_k(n)$ and $M_k(n)$ be the kth stage lower and upper bounds on x_n; we apply telescoping cancellations and elementary estimations sequenced thus: from $m_1(n)$ to $M_1(n)$ to $m_2(n)$ to $M_2(n)$ to ....

2. EULER'S IDENTITY AND VIETA'S FORMULA

Starting from the trigonometric formula

$$\sin y = 2 \sin \frac{y}{2} \cos \frac{y}{2}$$

we replace y successively by $x, x/2, x/2^2, \ldots, x/2^{n-1}$, multiply the results, and use the multiplicative form of telescoping cancellation, obtaining

$$\frac{\sin x}{2^n \sin \dfrac{x}{2^n}} = \prod_{j=1}^{n} \cos \frac{x}{2^j}.$$

Using the limit $\lim_{u \to 0}(\sin u / u) = 1$ as we have by letting n approach infinity Euler identity

$$\frac{\sin x}{x} = \prod_{j=1}^{\infty} \cos \frac{x}{2^j}. \tag{1}$$

The same process may be applied to the triplication formula $\sin 3y = 3 \sin y - 4 \sin^3 y$:

$$\frac{\sin y}{3 \sin \dfrac{y}{3}} = \frac{4 \cos^2 \dfrac{y}{3} - 1}{3};$$

proceeding as before, we find the analogue of (1)

$$\frac{\sin x}{x} = \prod_{j=1}^{\infty} \frac{4 \cos^2 \dfrac{x}{3^j} - 1}{3}.$$

Starting with (1), putting $x = \pi/2$, and evaluating the infinite product we obtain Vieta's formula, historically the first *exact* expression for π:

$$\frac{2}{\pi} = \frac{\sqrt{2}}{2}\left(\frac{\sqrt{2 + \sqrt{2}}}{2}\right)\left(\frac{\sqrt{2 + \sqrt{2 + \sqrt{2}}}}{2}\right)\cdots\;;$$

all that is necessary in this computation is the recursion $c_{n+1} = \sqrt{(1 + c_n)/2}$ for $c_n = \cos(\pi/2^n)$.

3. EULER'S PRODUCT AND CANTOR'S THEOREM

Using a familiar identity we have

$$\frac{1 - y^2}{1 - y} = 1 + y;$$

replacing y successively by x, x^2, x^{2^2}, ..., $x^{2^{n-1}}$, multiplying and canceling, we get

$$\frac{1 - x^{2^n}}{1 - x} = \prod_{j=0}^{n-1}(1 + x^{2^j}).$$

If $|x| < 1$ we may pass to the limit ($n \to \infty$) and we obtain the Euler product

$$\frac{1}{1 - x} = \prod_{n=0}^{\infty}(1 + x^{2^n}). \tag{1}$$

Since the left-hand side is the geometric series $1 + x + x^2 + \cdots$, we find by comparing the exponents of like powers that (1) is the analytic equivalent of a familiar fact in arithmetic: every positive integer is a unique sum of increasing powers of 2.

From (1) we may deduce the following theorem of Georg Cantor: every real number $A > 1$ may be uniquely represented as the product

$$A = \prod_{n=1}^{\infty}\left(1 + \frac{1}{a_n}\right)$$

where a_1, a_2, ... are positive integers satisfying the condition $a_{n+1} \geq a_n^2$. Moreover, A is rational if and only if n_0 exists such that $a_{n+1} = a_n^2$ for $n \geq n_0$.

We may assume without loss of generality that $1 < A \leq 2$ by replacing, if necessary, A with $2^s A_1$. In this case we have $a_1 = a_2 = \cdots = a_s = 1$. For $A \leq 2$ we put

$$a_1 = \left[\frac{A}{A - 1}\right], \qquad B = \frac{a_1 A}{a_1 + 1}, \qquad a_2 = \left[\frac{B}{B - 1}\right], \qquad C = \frac{a_2 B}{a_2 + 1}, \quad \text{etc.,}$$

where $[x]$ denotes as usual the greatest integer $\leq x$. It is now verified that $1 < B \leq 2$, $1 < C \leq 2$, ..., that $a_1{}^2 \leq a_2$, $a_2{}^2 \leq a_3$, ..., and that the a's are unique since

$$A < \prod_{n=0}^{\infty} (1 + a_1{}^{-2^n}) = \frac{a_1}{a_1 - 1}$$

by Euler's product; hence $a_1 \leq A/(A-1) < a_1 + 1$, etc.

Finally, if A is rational it may be shown that one of the numbers $A, B, C, \ldots$ must be of the form $K/(K-1)$ where K is an integer, and the remainder of the infinite product is then

$$\left(1 + \frac{1}{K}\right)\left(1 + \frac{1}{K^2}\right)\left(1 + \frac{1}{K^4}\right) \cdots = \frac{K}{K-1}.$$

From Cantor's theorem it follows that for every integer $a > 1$ the value of

$$\prod_{n=0}^{\infty} (1 - a^{-2^n})$$

is irrational since its reciprocal

$$\prod_{n=0}^{\infty} \left(1 + \frac{1}{a^{2^n} - 1}\right)$$

is irrational by Cantor's theorem.

Two related results due also to Cantor deal with a generalization of decimal and factorial expansions:

(a) If $0 < x \leq 1$ there exist unique positive integers $n_1, n_2, n_3, \ldots$ such that $1 < n_1 \leq n_2 \leq \cdots$ and

$$x = \frac{1}{n_1} + \frac{1}{n_1 n_2} + \frac{1}{n_1 n_2 n_3} + \cdots,$$

moreover, x is rational if and only if all n_k's are eventually equal;

(b) Let $a_1, a_2, a_3, \ldots$ be a sequence of positive integers such that for every positive integer n, $n | a_1 a_2 \cdots a_k$ for a suitable k, let $c_0, c_1, c_2, \ldots$ be non-negative integers satisfying $0 \leq c_n < a_n$ and $c_n \neq 0$ infinitely often, let

$$x = c_0 + \frac{c_1}{a_1} + \frac{c_2}{a_1 a_2} + \frac{c_3}{a_1 a_2 a_3} + \cdots$$

be a convergent series; then x is rational if and only if $c_n = a_n - 1$ for all sufficiently large n. By taking $n_i = i + 1$ in (a) we see that e is irrational. Using (b) we can prove the irrationality of such numbers as

$$\sum_{n=0}^{\infty} \frac{[\sqrt{n}]}{n!}, \qquad I_0(1) = 1 + \frac{1}{2^2} + \frac{1}{2^2 4^2} + \frac{1}{2^2 4^2 6^2} + \cdots,$$

and many others. For instance, the reader may wish to show that if $a_0, a_1, a_2, \ldots$ are nonnegative integers such that

$$f(x) = \sum_{'n=0}^{\infty} a_n x^n$$

is a non-polynomial rational function, then the number

$$a_0 + \frac{a_1}{1} + \frac{a_2}{1 \cdot 2^2} + \frac{a_3}{1 \cdot 2^2 3^3} + \cdots$$

is irrational.

4. TELESCOPING COINCIDENCE

Telescoping cancellation is loosely related to what might be called telescoping coincidence. To illustrate this we consider the series

$$f(z) = \sum_{n=-\infty}^{\infty} a_n(z) \quad \text{where} \quad a_n(z) = (z - n\pi)^{-2}.$$

The series is uniformly convergent on any compact set which excludes $z = \pm n\pi$, $n = 0, 1, \ldots$. If we consider $f(z + \pi)$ we find $a_n(z + \pi) = a_{n+1}(z)$; hence the series for $f(z + \pi)$ is the original series again, since shifting the terms back by one has no effect. It follows that $f(z)$ has period π. We could conclude this by recognizing $f(z)$ as the Mittag–Leffler expansion of $\operatorname{cosec}^2 z$ but we needn't know this to infer periodicity. Similar phenomenon occurs in the construction of the Weierstrass function

$$p(z) = \sum_{-\infty}^{\infty} \sum_{-\infty}^{\infty} [(z + 2m\omega_1 + 2n\omega_2)^{-2} - (2m\omega_1 + 2n\omega_2)^{-2}]$$

where ω_1 and ω_2 are two complex numbers whose ratio is not real; the summation extends over all integers except for $m = n = 0$ where the term $(2m\omega_1 + 2n\omega_2)^{-2}$ is left out. When z is replaced by $z + 2\omega_1$ or by $z + 2\omega_2$, we find similar coincidence of original and shifted terms; hence $p(z)$ is doubly periodic with periods $2\omega_1$ and $2\omega_2$. The arrangement of factors inside the square brackets ensures uniform convergence (except at $z = 2n\omega_1 + 2m\omega_2$) of the defining series.

Related use of incomplete telescoping coincidence occurs in the following. To show that the function

$$f(z) = \sum_{n=0}^{\infty} z^{2^n} \tag{1}$$

has the unit circle as its natural boundary we observe what happens on replacing z by z^2; since $z^{2^{n+1}} = (z^{2^n})^2$ we have

$$f(z^2) = -z + f(z). \tag{2}$$

Since $f(z)$ obviously has a singularity at $z = 1$ it follows from (2) that there are singularities at the two second roots of 1, also at the four fourth roots, the eight eighth roots, etc. Hence the singularities are dense on $|z| = 1$ which is therefore a natural boundary.

A more recondite use of this type of telescoping coincidence was made by Poincaré and others in constructing functions invariant under an infinite discrete group G of transformations. This invariance generalizes the concept of periodicity: if $f(x)$ is periodic with period a, then G is the group of translations generated by the shift $x \to x + a$. In the general case it is observed that if the series

$$f(z) = \sum_{g \in G} \phi(gz)$$

converges, then by the definition of a group $f(z) = f(gz)$ for every element g of G—hence the required invariance.

5. SUMMATION OF CERTAIN SERIES

Suppose that a (finite or infinite) power series has a known sum:

$$\sum_{n=0}^{\infty} a_n x^n = f(x) \tag{1}$$

and let $P(n)$ be a polynomial, then it is possible to evaluate the sum of the series

$$\sum_{n=0}^{\infty} a_n P(n) x^n.$$

We may assume by linearity that $P(n) = n^k$, let

$$\sum_{n=0}^{\infty} a_n n^k x^n = G(x).$$

Starting with (1) we differentiate-and-multiply-by-x k times to obtain

$$G(x) = \left(x \frac{d}{dx} \right)^k f(x).$$

This may be written as

$$G(x) = \sum_{j=0}^{k} S(k,j) x^j f^{(j)}(x)$$

where $S(k,j)$ is the Stirling number of the second kind [61]. Often a specific series

$$a_0 + 1^k a_1 + 2^k a_2 + \cdots$$

can be summed by introducing a parameter x to form the power series $a_0 + a_1 x + a_2 x^2 + \cdots$, applying the above, and putting $x = 1$. For instance, to obtain the mean and variance of the binomial distribution it is necessary to evaluate the sums

$$S_1 = \sum_{k=0}^{n} k \binom{n}{k}, \qquad S_2 = \sum_{k=0}^{n} k^2 \binom{n}{k}.$$

The binomial theorem gives us immediately

$$S_1 = x \frac{d}{dx} (1 + x)^n \bigg|_{x=1} = 2^{n-1} n,$$

$$S_2 = \left(x \frac{d}{dx} \right)^2 (1 + x)^n \bigg|_{x=1} = 2^{n-2} n(n + 1).$$

Somewhat more generally, to evaluate

$$S = \sum_{k=0}^{n} \binom{n}{k} k^p (n - k)^q u^k v^{n-k}$$

we start with

$$f(x, y, u, v) = \sum_{k=0}^{n} \binom{n}{k} (ux)^k (vy)^{n-k} = (ux + vy)^n$$

and we have

$$S = \left(x \frac{\partial}{\partial x} \right)^p \left(y \frac{\partial}{\partial y} \right)^q (ux + vy)^n \bigg|_{x=y=1}.$$

Again, let

$$\sum_{n=1}^{\infty} a_n x^n = f(x) \tag{2}$$

be known and let I be the operator inverse to $x(d/dx)$:

$$If(x) = \int_0^x f(x) \frac{dx}{x}.$$

Then applying I to f k times we find

$$\sum_{n=1}^{\infty} a_n n^{-k} x^n = I^k f(x). \tag{3}$$

More generally, let $f(x)$ in (2) be known and consider the sum

$$\sum_{n=1}^{\infty} a_n \frac{P(n)}{Q(n)} x^n = H(x).$$

Using the decomposition into partial fractions we have

$$\frac{P(n)}{Q(n)} = P_1(n) + \sum_i \sum_k A_{ik}(n + b_i)^{-k}$$

where P_1 is a polynomial and A_{ik} and b_i are constants. To evaluate $H(x)$ it is enough to consider the sum

$$F(x) = \sum_{n=1}^{\infty} a_n(n + b_i)^{-k}x^n.$$

We apply a modification of (3):

$$\sum_{n=1}^{\infty} a_n x^{n+b} = x^b f(x), \qquad \sum_{n=1}^{\infty} a_n(n + b)^{-k}x^n = x^{-b}I^k x^b f(x).$$

The reader may observe connections with the previously mentioned digamma function and with the dilogarithms (which appear in the chapter on integrals).

Summation of certain series depends on filtering out an arithmetic subsequence of indices; this process is also known as multisection of series. Suppose that we start again with a power series

$$\sum_{n=0}^{\infty} a_n x^n = f(x)$$

whose sum $f(x)$ is known. Let p and q be integers with $p \geq 1$, $p + q \geq 0$ and suppose that we wish to evaluate

$$\sum_{n=0}^{\infty} a_{pn+q} x^{pn+q} = F(x)$$

or

$$\sum_{n=0}^{\infty} a_{pn+q} x^n = G(x).$$

We have

$$G(x) = x^{-q/p} F(x^{1/p});$$

to evaluate $F(x)$ we use the roots of unity: let $t = e^{2\pi i/p}$, then

$$T = \sum_{k=0}^{p-1} t^{-qk} t^{nk} = \sum_{k=0}^{p-1} t^{k(n-q)}$$

may be evaluated, being a geometric series. If $p \mid n - q$ each term is 1 and so $T = p$. If $p \nmid n - q$ then

$$T = \frac{1 - e^{2\pi i(n-q)}}{1 - e^{2\pi i(n-q)/p}}$$

so that $T = 0$. Hence

$$F(x) = \sum_{n=0}^{\infty} a_{pn+q} x^{pn+q} = \frac{1}{p} \sum_{k=0}^{p-1} t^{-qk} f(xt^k).$$

For instance, let

$$S = \binom{n}{1} + \binom{n}{4} + \cdots = \sum_{k=0}^{(n-1)/3} \binom{n}{3k+1};$$

we use the foregoing with $f(x) = (1 + x)^n$, $q = 1$, $p = 3$, and $t = e^{2\pi i/3}$ to obtain

$$\sum_{k=0}^{(n-1)/3} \binom{n}{3k+1} x^{3k+1} = \frac{1}{3} [(1 + x)^n + t^2 (1 + tx)^n + t(1 + t^2 x)^n];$$

putting $x = 1$ and observing that $1 + t + t^2 = 0$, we have

$$S = \frac{1}{3} (2^n + (-1)^n \varepsilon_n)$$

where $\varepsilon_n = -1$ if $n \equiv 0$ or $1 \pmod 3$, and $\varepsilon_n = 2$ if $n \equiv 2 \pmod 3$. This may also be written as

$$S = \frac{1}{3} \left(2^n + 2 \cos \frac{(n - 2)\pi}{3} \right).$$

Similarly, let

$$S(x) = \sum_{n=0}^{\infty} \frac{x^{4n}}{(4n)!};$$

we use the foregoing with $f(x) = e^x$, $q = 0$, $p = 4$, $t = i$, and we find

$$S(x) = \frac{1}{4} (e^x + e^{ix} + e^{-x} + e^{-ix}) = \frac{1}{2} (\cosh x + \cos x).$$

For a variety of combinatorial applications of the multisection technique see [62].

Certain series of the form $\sum_{n=1}^{\infty} f(n)$ and $\sum_{n=1}^{\infty} (-1)^n g(n)$ may be summed by an application of the residue theorem. Consider the integrals

$$I_n = \frac{1}{2\pi i} \int_{C_n} \pi f(z) \cot \pi z \, dz, \qquad J_n = \frac{1}{2\pi i} \int_{C_n} \pi g(x) \operatorname{cosec} \pi z \, dz$$

where the contour C_n is the square with the vertices $[n + (1/2)](\pm 1 \pm i)$, and the functions f and g are single-valued and analytic except for a finite number of non-integer poles; we require also that for $|z| = R$ large both $|f|$ and $|g|$ are of the order R^{-1-a}, $a > 0$. Since $\cot \pi z$ and $\operatorname{cosec} \pi z$ are bounded on C_n we use the elementary estimate

$$\left| \int_C F(z)\, dz \right| \le \text{Length } (C) \, \underset{C}{\text{Max}} \, |F(z)|$$

to conclude that $\lim I_n = \lim J_n = 0$ as $n \to \infty$. Therefore in either case the sum of all residues is 0. Since $\pi \cot \pi z$ and $\pi \, \text{cosec} \, \pi z$ have simple poles at all integers n, with residues 1 and $(-1)^n$ respectively, we find that

$$\sum_{n=-\infty}^{\infty} f(n) = -\pi \sum \text{Res}[f(z)\cot \pi z],$$

$$\sum_{n=-\infty}^{\infty} (-1)^n g(n) = -\pi \sum \text{Res}[g(z)\text{cosec } \pi z].$$

For instance, with $f(z) = (z^2 + a^2)^{-1}$, a real, we have

$$\sum_{n=-\infty}^{\infty} \frac{1}{n^2 + a^2} = \frac{1}{a^2} + 2\sum_{n=1}^{\infty} \frac{1}{n^2 + a^2} = \frac{\pi}{a} \coth \pi a.$$

6. PRODUCTS AND FACTORIZATION

To evaluate the product

$$P = \prod_{j=1}^{m-1} \sin \frac{j\pi}{m}$$

we start in a rather unobvious way by proving the factorization

$$r^{2m} - 2r^m \cos m\theta + 1 = \prod_{k=0}^{m-1} \left[r^2 - 2r\cos\left(\theta + \frac{2k\pi}{m}\right) + 1 \right]. \tag{1}$$

Let $z = re^{i\theta}$, $\bar{z} = re^{-i\theta}$, $t = e^{2\pi i/m}$; then we have the cyclotomic products

$$z^m - 1 = \prod_{k=0}^{m-1} (z - t^k), \qquad \bar{z}^m - 1 = \prod_{k=0}^{m-1} (\bar{z} - t^k)$$

and by *suitable pairing* of linear factors

$$(z^m - 1)(\bar{z}^m - 1) = \prod_{k=0}^{m-1} (z - t^k)(\bar{z} - t^{m-k}). \tag{2}$$

By DeMoivre's theorem the left-hand sides of (1) and (2) are equal. Further

$$(z - t^k)(\bar{z} - t^{m-k}) = r^2 - 2r\cos\left(\theta - \frac{2k\pi}{m}\right) + 1$$

and replacing θ by $-\theta$ we have (1). An alternative way to prove (1) is to show first, by induction for instance, that the quadratic polynomial in r, $r^2 - 2r\cos\theta + 1$, divides the polynomial $r^{2m} - 2r^m \cos m\theta + 1$ of degree $2m$ in r; then we observe that the latter is unchanged if θ is replaced by $\theta + (2k\pi/m)$, so that

$r^2 - 2r\cos[\theta + (2k\pi/m)] + 1$ is also a factor; the rest follows by comparing constant terms.

In (1) we put $r = 1$ and $\theta = 2\phi$, using the identity $1 - \cos 2x = 2\sin^2 x$ we obtain

$$\frac{\sin m\phi}{\sin \phi} = 2^{m-1} \prod_{k=1}^{m-1} \sin\left(\phi + \frac{k\pi}{m}\right)$$

and we let $\phi \to 0$ using $\lim \sin x/x = 1$ as $x \to 0$, to get

$$P = \prod_{k=1}^{m-1} \sin \frac{k\pi}{m} = 2^{1-m}m.$$

Several other occasionally useful factorizations may be similarly obtained:

$$\cos n\theta - \cos n\beta = 2^{n-1} \prod_{k=0}^{n-1} \left[\cos \theta - \cos\left(\beta + \frac{2k\pi}{n}\right)\right],$$

$$\cos n\theta \quad\begin{cases} = 2^{n-1} \cos \theta \prod_{k=0}^{(n-3)/2} \left[\cos^2 \theta - \cos^2 \frac{(2k+1)\pi}{2n}\right], & n \text{ odd} \\[3mm] = 2^{n-1} \prod_{k=0}^{(n/2)-1} \left[\cos^2 \phi - \cos^2 \frac{(2k+1)\pi}{2n}\right], & n \text{ even} \end{cases}$$

$$\frac{\sin n\theta}{n \sin \theta} \quad\begin{cases} = \prod_{k=1}^{(n-1)/2} \left(1 - \frac{\sin^2 \theta}{\sin^2 \frac{k\pi}{n}}\right), & n \text{ odd} \\[5mm] = \cos \theta \prod_{k=1}^{(n-2)/2} \left(1 - \frac{\sin^2 \theta}{\sin^2 \frac{k\pi}{n}}\right), & n \text{ even} \end{cases}$$

Considering either of the last two equations we may take the limit as $\theta \to 0$, $n \to \infty$, $n\theta \to x$; with some precautions as to convergence we obtain the infinite product

$$\sin x = x \prod_{k=1}^{\infty} \left(1 - \frac{x^2}{k^2\pi^2}\right).$$

An alternative statement of this is

$$\frac{\sin \pi x}{\pi x} = \prod_{n=1}^{\infty} \left(1 - \frac{x^2}{n^2}\right). \tag{3}$$

The following method of obtaining the infinite product for $\sin x/x$ is due to Darboux. We have

$$\frac{\sin x}{x} = \frac{e^{ix} - e^{-ix}}{2ix} = \frac{1}{2ix} \lim_{n\to\infty} \left[\left(1 + \frac{ix}{n}\right)^n - \left(1 - \frac{ix}{n}\right)^n\right].$$

Put

$$F_n(x) = \frac{1}{2i}\left[\left(1 + \frac{ix}{n}\right)^n - \left(1 - \frac{ix}{n}\right)^n\right]$$

and let $x = n \tan \theta$, then

$$F_n(x) = \frac{1}{2i} \sec{}^n\theta(e^{in\theta} - e^{-in\theta})$$

which vanishes if and only if $e^{2in\theta} = 1$ or $\theta = \pm k\pi/n$. It follows that $x \pm n \tan k\pi/n$ divides $F_n(x)$, and since $F_n(x)$ is a polynomial of degree n in x, we find

$$F_n(x) = Ax \prod_{k=1}^{(n-1)/2}\left(1 - \frac{x^2}{n^2 \tan^2 \dfrac{k\pi}{n}}\right) \qquad \text{for } n \text{ odd.} \qquad (4)$$

Here A is a constant and we find easily that $A = 1$. Now, taking the limit in (4) as $n \to \infty$ we obtain (3) again.

7. INFINITE PRODUCTS AND Γ-FUNCTION

Infinite products of the type

$$P = \prod_{n=1}^{\infty} \frac{P(n)}{Q(n)}$$

can be evaluated in terms of Γ-functions provided that P and Q are polynomials which have: (a) the same degree, (b) the same two leading coefficients, (c) no zeros which are positive integers. For then we have, by factorizing,

$$P = \prod_{n=1}^{\infty} \frac{(n - a_1)(n - a_2) \cdots (n - a_k)}{(n - b_1)(n - b_2) \cdots (n - b_k)}$$

where $\sum a_j = \sum b_j$; using the Weierstrass product

$$\frac{1}{\Gamma(x)} = xe^{\gamma x} \prod_{n=1}^{\infty} \left(1 + \frac{x}{n}\right)e^{-x/n}$$

and the recursion $\Gamma(x + 1) = x\Gamma(x)$, we find that

$$P = \prod_{j=1}^{k} \frac{\Gamma(1 - b_j)}{\Gamma(1 - a_j)}.$$

For instance, taking $a_1 = x$, $a_2 = -x$, $b_1 = b_2 = 0$, where x is not an integer, and using (3) of the last section, we have

$$x \prod_{n=1}^{\infty} \left(1 - \frac{x^2}{n^2}\right) = \frac{\sin \pi x}{\pi} = \frac{1}{\Gamma(x)\Gamma(1 - x)}. \tag{1}$$

Putting here $x = 1/2$ we find $\Gamma(1/2) = \sqrt{\pi}$ and the Wallis product

$$\frac{\pi}{2} = \prod_{n=1}^{\infty} \frac{(2n)(2n)}{(2n - 1)(2n + 1)}.$$

Replacing x in (1) by iy we find

$$\prod_{n=1}^{\infty} \left(1 + \frac{y^2}{n^2}\right) = \frac{\sinh \pi y}{\pi y}$$

and in particular with $y = 1$

$$\prod_{n=1}^{\infty} \left(1 + \frac{1}{n^2}\right) = \frac{e^{\pi} - e^{-\pi}}{2\pi}.$$

Taking in the basic product $a_1 = x$, $a_2 = -x$, $a_3 = ix$, $a_4 = -ix$ (x not an integer) and $b_1 = b_2 = b_3 = b_4 = 0$ we have

$$\prod_{n=1}^{\infty} \left(1 - \frac{x^4}{n^4}\right) = \frac{\sin \pi x \sinh \pi x}{\pi^2 x^2}.$$

Replacing here x by

$$y\sqrt{i} = 2^{-1/2} y(1 + i)$$

we have

$$\prod_{n=1}^{\infty} (1 + y^4/n^4) = \left(\sin^2 \frac{\pi y}{\sqrt{2}} + \sinh^2 \frac{\pi y}{\sqrt{2}}\right) \Big/ \pi^2 y^2.$$

In particular, putting $y = 1$ and $y = p\sqrt{2}$ (where p is an integer) we have

$$\prod_{n=1}^{\infty} (1 + n^{-4}) = \pi^{-2} \left(\sin^2 \frac{\pi}{\sqrt{2}} + \sinh^2 \frac{\pi}{\sqrt{2}}\right),$$

$$\prod_{n=1}^{\infty} (1 + 4p^4/n^4) = (\sinh^2 p\pi)/2p^2\pi^2.$$

The following infinite product, due to Mellin [55], is similarly obtained: let

$$P(x) = a_1 x + a_2 x^2 + \cdots + a_n x^n,$$

$$x^n[1 + P(1/x)] = \prod_{k=1}^{n} (x - r_k),$$

then

$$\prod_{n=0}^{\infty} \left[1 + P\left(\frac{y}{x+n}\right)\right] e^{-a_1 y/n} = e^{-\gamma a_1 y} \Gamma^n(x) \Big/ \prod_{k=1}^{n} \Gamma(x - r_k y).$$

From this we may deduce

$$\prod_{n=2}^{\infty} \frac{n^3 - 1}{n^3 + 1} = \frac{2}{3} \qquad \text{(Gram's product)},$$

$$\prod_{n=-\infty}^{\infty} \left[1 - \frac{y^2}{(n+x)^2}\right] = 1 - \frac{\sin^2 \pi y}{\sin^2 \pi x} \qquad \text{(Euler's product)}.$$

Two infinite products of a different type are

$$\frac{\pi}{2e} = \prod_{n=1}^{\infty} \left(1 + \frac{2}{n}\right)^{(-1)^n n}, \tag{2}$$

$$\frac{6}{\pi e} = \prod_{n=2}^{\infty} \left(1 + \frac{2}{n}\right)^{(-1)^n n}. \tag{3}$$

Here for convergence we take factors in pairs:

$$P = \prod_{1}^{\infty} p_n = \lim_{n \to \infty} (p_1 p_2)(p_3 p_4) \cdots (p_{2n-1} p_{2n}) \cdots.$$

(2) and (3) are of a rather unlikely provenance. Let B_n be the ball of radius R in the n-dimensional Euclidean space, that is,

$$B_n = \{(x_1, \ldots, x_n) : x_1^2 + \cdots + x_n^2 \le R^2\};$$

let c_n be a cylinder inscribed into B_n, of base-radius r and height $2h$, that is,

$$c_n = \{(x_1, \ldots, x_n) : x_1^2 + \cdots + x_{n-1}^2 \le r^2, -h \le x_n \le h, h^2 + r^2 = R^2\}.$$

Let C_n be the inscribed cylinder of largest possible volume. Using the formula for the volume of B_n:

$$\text{Vol}(B_n) = \pi^{n/2} R^n / \Gamma(1 + n/2)$$

and the property of the Γ-function

$$\lim_{x \to \infty} \frac{\Gamma(x + a)}{x^a \Gamma(x)} = 1,$$

it is shown that if we put

$$p_n = \text{Vol}(C_n)/\text{Vol}(B_n)$$

then

$$p = \lim_{n \to \infty} p_n = (2/\pi e)^{1/2}. \tag{4}$$

Next, introducing the ratio $s_n = p_{n+2}/p_n$ and applying the recursion $\Gamma(x + 1) = x\Gamma(x)$, we find

$$s_n = \left(\frac{n}{n+2}\right)^{n/2} \left(\frac{n+1}{n-1}\right)^{(n-1)/2} \tag{5}$$

By an easy calculation $p_2 = 2/\pi$ and $p_3 = 3^{-1/2}$ and by the telescoping cancellation

$$p = p_2 \frac{p_4}{p_2} \frac{p_6}{p_4} \cdots = \frac{2}{\pi} \prod_{n=1}^{\infty} s_{2n},$$

$$p = p_3 \frac{p_5}{p_3} \frac{p_7}{p_5} \cdots = 3^{-1/2} \prod_{n=1}^{\infty} s_{2n-1}.$$

Now, applying (4) and (5) and squaring, we obtain (2) and (3).

8. THE McLAURIN SERIES FOR tan x AND sec x

Power series for $\sin x$ and $\cos x$ are well known and easy to obtain, those for $\tan x$ and $\sec x$ are neither. A direct attempt to obtain the power series for $\tan x$ based on the definition $\tan x = \sin x/\cos x$ requires the division of two power series; the inverse function arc $\tan x$ has a simple power series but to get from it the series for $\tan x$ requires the inversion of a power series; the McLaurin formula requires the value of the nth derivative of $\sin x/\cos x$ at $x = 0$. While each of these methods could be pushed through, though not very easily, to yield the desired series, we would still be rather ignorant of the nature of the coefficients in them as compared to, say, the factorials in the series for $\sin x$ and $\cos x$.

The present section belongs really to the subject of combinatorics, under generating functions, but it is included here for formal reasons; we shall find the expansions

$$\tan x = A_1 x + \frac{A_3}{3!} x^3 + \frac{A_5}{5!} x^5 + \cdots \tag{1}$$

$$\sec x = A_0 + \frac{A_2}{2!} x^2 + \frac{A_4}{4!} x^4 + \cdots \tag{2}$$

not by starting with the two functions and processing them, but by introducing certain combinatorial quantities $A_0, A_1, \ldots$, setting up their so-called exponential generating function

$$y(x) = \sum_{n=0}^{\infty} \frac{A_n}{n!} x^n$$

and proving that $y(x) = \tan(\pi/4 + x/2)$ from which (1) and (2) follow at once. We obtain $y(x)$ by showing that, roughly speaking, the combinatorial rules of formation of A_n's force $y(x)$ to satisfy the same differential equation as $\tan x$.

A permutation $(k_1, k_2, \ldots, k_n)$ of $(1, 2, \ldots, n)$ is called a zigzag if no three neighbors k_i, k_{i+1}, k_{i+2} $(i = 1, 2, \ldots, n - 2)$ are in the natural order: for no such i is $k_i < k_{i+1} < k_{i+2}$ or $k_{i+2} < k_{i+1} < k_i$. Thus (31425) is a zigzag and (34125) is not. We observe that if $(k_1, k_2, \ldots, k_n)$ is a zigzag then so is its complement $(n + 1 - k_1, n + 1 - k_2, \ldots, n + 1 - k_n)$, hence:

(a) There are as many initially rising zigzags $(k_1 < k_2)$ as falling ones $(k_1 > k_2)$.
(b) There are as many finally rising zigzags $(k_{n-1} < k_n)$ as finally falling ones $(k_{n-1} > k_n)$.
(c) The number $2A_n$ of all zigzags is even.

To be able to count the number $2A_n$ of all n-zigzags we introduce *some further structure*. Consider a zigzag in which the largest number n occupies the $(r + 1)$th place: $k_{r+1} = n$. Then any zigzag on the preceding r elements, which is finally falling, and any zigzag on the $n - r - 1$ following elements, which is initially rising, may be combined, together with n in the $(r + 1)$th place, to produce a zigzag on all n elements. Conversely, any n-zigzag can be so decomposed. To simplify matters we define $A_0 = A_1 = A_2 = 1$, then the above argument shows that

$$2A_n = \sum_{r=0}^{n-1} \binom{n-1}{r} A_r A_{n-1-r},$$

the binomial coefficient $\binom{n-1}{r}$ counting the ways of splitting the $n - 1$ numbers $1, 2, \ldots, n - 1$ into two sets of sizes r and $n - 1 - r$. Let $A_j = j! a_j$ then

$$2n a_n = \sum_{r=0}^{n-1} a_r a_{n-1-r}. \tag{3}$$

If now we introduce the generating function

$$y(x) = a_0 + a_1 x + a_2 x^2 + \cdots$$

then the recurrence (3) becomes

$$1 + y^2 = 2y' \quad \text{or} \quad \frac{y'}{1 + y^2} = \frac{1}{2}.$$

Integrating and noting that $y(0) = 1$ we have

$$y(x) = \tan\left(\frac{\pi}{4} + \frac{x}{2}\right), \quad y(-x) = \tan\left(\frac{\pi}{4} - \frac{x}{2}\right)$$

and by adding and subtracting these

$$\sec x = a_0 + a_2 x^2 + a_4 x^4 + \cdots \tag{4}$$

$$\tan x = a_1 x + a_3 x^3 + \cdots . \tag{5}$$

The recursion (3) together with the initial values $a_0 = a_1 = 1$, $a_2 = 1/2$, allows us to calculate the A's and a's:

n	0	1	2	3	4	5	6	7	8
a_n	1	1	1/2	1/3	5/24	2/15	61/720	17/315	277/8064
A_n	1	1	1	2	5	16	61	272	1385.

With some elements of complex variables we find that the series (4) and (5) converge for $|x| < \pi/2$; further

$$\lim_{n \to \infty} \left(\frac{A_n}{n!} \right)^{1/n} = \frac{2}{\pi} .$$

The method given here is due to D. André [1] who introduced the concept of a zigzag permutation. This is also known sometimes as an alternating permutation; the latter is not a good name since it may bring some confusion with the even permutations which make up the alternating group (see also [20]).

9. EULER ACCELERATION AND THE TAYLOR SERIES

In this section we start to use the symbol-reification principle. One might put it simply and less precisely by saying that we start to use operatorial and symbolic methods. The principle in question might be said to consist of transferring to operations the properties of quantities.

Accordingly, the results are tentative and require either a rigorous justification of the method or separate verification in each specific case. However, it is usually easier to prove that a right-hand side equals a left-hand side than to start with one side only. As our first example we consider the Euler convergence-acceleration. Suppose that a slowly convergent series

$$S = a_1 - a_2 + a_3 - \cdots$$

is to be evaluated, e.g.,

$$\log 2 = 1 - \frac{1}{2} + \frac{1}{3} - \cdots .$$

We write it symbolically as

$$S = (1 - E + E^2 - \cdots)a_1 = \frac{1}{1 + E} a_1$$

where E is the shift operator: $Ea_n = a_{n+1}$. Let also Δ be the difference operator: $\Delta a_n = a_{n+1} - a_n$, $\Delta^2 a_n = \Delta(\Delta a_n) = a_{n+2} - 2a_{n+1} + a_n$, etc. We have the symbolic identity $E = 1 + \Delta$ so that

$$S = \frac{1}{2 + \Delta} a_1 = \frac{1}{2} \frac{1}{1 + \dfrac{\Delta}{2}} a_1 = \left(\frac{1}{2} - \frac{\Delta}{2^2} + \frac{\Delta^2}{2^3} - \cdots \right) a_1$$

by the geometric series expansion. This gives us Euler's formula

$$S = \frac{a_1}{2} - \frac{1}{2^2} \Delta a_1 + \frac{1}{2^3} \Delta^2 a_1 - \cdots. \tag{1}$$

In our example $\log 2 = 1 - (1/2) + (1/3) - \cdots$ so that $a_n = 1/n$; the higher differences may be computed by induction on k: $\Delta^k a_n = (-1)^k \, k!/n(n+1) \cdots (n+k)$ whence

$$\Delta^k a_1 = (-1)^k/(k+1).$$

Therefore the transformed series is

$$\log 2 = \sum_{k=1}^{\infty} \frac{1}{2^k k}$$

showing spectacularly faster convergence. In fact, unlike the original series, this one is practically convergent, not only convergent (i.e., it may be used for computation to reasonable accuracy in reasonable time).

It may be checked that the Euler method transforms the geometric series $1 - b + b^2 - b^3 + \cdots$ into the geometric series

$$\frac{1}{2}\left[1 - \left(\frac{b-1}{2} \right) + \left(\frac{b-1}{2} \right)^2 - \cdots \right],$$

hence the convergence actually deteriorates if $b < 1/3$. However, if the original series converges, so does the transformed one.

Similar symbolic procedures may be applied to power series. For instance,

$$a_0 - a_1 x + a_2 x^2 - \cdots = (1 - xE + x^2 E^2 - \cdots)a_0$$

$$= \frac{1}{1 + xE} a_0 = \frac{1}{1 + x + x\Delta} a_0$$

$$= \frac{1}{1 + x} \frac{1}{1 + \dfrac{x}{1 + x} \Delta} a_0$$

$$= \frac{a_0}{1 + x} - \frac{x}{(1 + x)^2} \Delta a_0 + \frac{x^2}{(1 + x)^3} \Delta^2 a_0 - \cdots.$$

Similarly,

$$a_1 x + a_2 x^3 + a_3 x^5 + \cdots$$

$$= \frac{x}{1 - x^2 E} a_1 = \frac{x}{1 - x^2} \frac{1}{1 - \dfrac{x^2}{1 - x^2} \Delta} a_1$$

$$= \sqrt{1 + y^2}(y a_1 + y^3 \Delta a_1 + y^5 \Delta^2 a_1 + \cdots), \qquad y = \frac{x}{\sqrt{1 - x^2}},$$

and

$$a_1 x - a_2 x^3 + a_3 x^5 - \cdots$$

$$= \frac{x}{1 + x^2 E} a_1 = \frac{x}{1 + x^2} \frac{1}{1 + \dfrac{x^2}{1 + x^2} \Delta} a_1$$

$$= \sqrt{1 - y^2}[y a_1 - y^3 \Delta a_1 + y^5 \Delta^2 a_1 - \cdots], \qquad y = \frac{x}{\sqrt{1 + x^2}}.$$

In particular, with $a_1 = 1, a_2 = 1/3, a_3 = 1/5$, etc., the last two formulas become

$$\frac{1}{\sqrt{1 + y^2}} \log(y + \sqrt{1 + y^2}) = y - \frac{2}{3} y^3 + \frac{2 \cdot 4}{3 \cdot 5} y^5 - \frac{2 \cdot 4 \cdot 6}{3 \cdot 5 \cdot 7} y^7 + \cdots,$$

$$\frac{1}{\sqrt{1 - y^2}} \arcsin y = y + \frac{2}{3} y^3 + \frac{2 \cdot 4}{3 \cdot 5} y^5 + \frac{2 \cdot 4 \cdot 6}{3 \cdot 5 \cdot 7} + y^7 \cdots.$$

Integrating these with respect to y from 0 to x and using the integration by parts we find

$$\frac{1}{2} [\operatorname{arg\,sinh} x]^2 = \frac{x^2}{2} - \frac{2}{3} \frac{x^4}{4} + \frac{2 \cdot 4}{3 \cdot 5} \frac{x^6}{6} - \frac{2 \cdot 4 \cdot 6}{3 \cdot 5 \cdot 7} \frac{x^8}{8} + \cdots$$

$$\frac{1}{2} [\arcsin x]^2 = \frac{x^2}{2} + \frac{2}{3} \frac{x^4}{4} + \frac{2 \cdot 4}{3 \cdot 5} \frac{x^6}{6} + \frac{2 \cdot 4 \cdot 6}{3 \cdot 5 \cdot 7} \frac{x^8}{8} + \cdots.$$

As another example we consider the Taylor series for function $f(x)$ of one variable:

$$f(x + a) = f(x) + \frac{f'(x)}{1!} a + \frac{f''(x)}{2!} a^2 + \cdots; \tag{2}$$

let E_a and D stand for the shift and the differentiator: $E_a f(x) = f(x + a)$, $Df(x) = f'(x)$. Then (2) may be written symbolically as

$$E_a = e^{aD} \qquad (3)$$

where the expansion of e^{aD} is obtained by using the exponential series. Similarly, the Taylor theorem for a function $f(x, y)$ of two variables is

$$f(x + h, y + k) = f(x, y) + \frac{1}{1!}\,[hf_x(x, y) + kf_y(x, y)]$$

$$+ \frac{1}{2!}\,[h^2 f_{xx}(x, y) + 2hk f_{yx}(x, y) + k^2 f_{yy}(x, y)] + \cdots$$

which may be written symbolically as

$$E_{hk} = e^{h(\delta/\delta x) + k(\delta/\delta y)}; \qquad (4)$$

here E_{hk} is the shift operator: $E_{hk} f(x, y) = f(x + h, y + k)$, and the right-hand side is obtained by using the exponential series and then the binomial theorem.

The mnemonic and suggestive value of symbolisms (3) and (4) is undoubted but the reader may wonder what sort of mathematical foundation, if any, is behind them. It turns out that such a foundation is found in the theory of semigroups of operators [31]. The shifts E_a form a semigroup (here even a group): $E_a E_a f(x) = E_{a+b} f(x) = f(x + a + b)$. The differentiator D is then what is called the infinitesimal generator of the semigroup. The idea here is that a semigroup of transformations $\{T_s\}$, defined by the relation $T_s T_r = T_{s+r}$, behaves like the special exponential semigroup $\{e^{as}\}$ on account of the index law $e^{as} e^{ar} = e^{a(s+r)}$. In the exponential case a could be determined from the obvious relation

$$\left.\frac{de^{as}}{ds}\right|_{s=0} = a;$$

in the general case the equivalent of a, i.e., the infinitesimal generator, may be defined by the corresponding limit

$$\lim_{s \to 0} \frac{T_s - I}{s} \qquad (5)$$

where I is the identity operator, if this limit exists. When T_s is the shift operator we have

$$\frac{T_s - I}{s} f(x) = \frac{f(x + s) - f(x)}{s}$$

and hence

$$\lim_{s \to 0} \frac{T_s - I}{s} = D$$

where D is the differentiator.

An interesting example is the Dirichlet semigroup $\{D_s\}$ of operators acting on power series

$$f(x) = \sum_{n=1}^{\infty} a_n x^n$$

which are suitably convergent and have no constant term:

$$D_s \sum_{n=1}^{\infty} a_n x^n = \sum_{n=1}^{\infty} a_n n^{-s} x^n.$$

We verify the index law $D_s D_s = D_{r+r}$ and the integral representation

$$D_s f(x) = \frac{1}{\Gamma(s)} \int_0^{\infty} z^{s-1} f(xe^{-z})\, dz. \tag{6}$$

By taking $f(x) = x/(1 - x)$, applying (6), and then putting $x = 1$, we have Riemann's formula for the ζ-function:

$$\zeta(s) = \frac{1}{\Gamma(s)} \int_0^{\infty} \frac{z^{s-1}}{e^z - 1}\, dz, \qquad \mathrm{Re}(s) > 0.$$

Making use of the semigroup property, we obtain a hierarchy of Riemann-type formulas:

$$\zeta(s_1 + s_2) = \frac{1}{\Gamma(s_1)\Gamma(s_2)} \int_0^{\infty} \int_0^{\infty} \frac{z_1^{s_1-1} z_2^{s_2-1}}{e^{z_1 + z_2} - 1}\, dz_1\, dz_2, \tag{7}$$

$$\zeta(s_1 + s_2 + s_3) = \frac{1}{\Gamma(s_1)\Gamma(s_2)\Gamma(s_3)} \int_0^{\infty} \int_0^{\infty} \int_0^{\infty} \frac{z_1^{s_1-1} z_2^{s_2-1} z_3^{s_3-1}}{e^{z_1 + z_2 + z_3} - 1}\, dz_1\, dz_2\, dz_3, \tag{8}$$

and so on. Certain rather complicated integrals can be simply obtained by transforming in various ways the multiple integrals in (7) and (8). For instance, if we change the variables z_1 and z_2 in (7) to t and v by $z_1 = tv$, $z_1 + z_2 = v$, we have the formula for the Beta function

$$B(s_1, s_2) = \int_0^1 t^{s_1-1}(1 - t)^{s_2-1}\, dt = \frac{\Gamma(s_1)\Gamma(s_2)}{\Gamma(s_1 + s_2)}.$$

If, instead, we use the polar coordinates r and ϕ: $z_1 = r \cos \phi$, $z_2 = r \sin \phi$, we have another representation for the Beta function:

$$\frac{\Gamma(s_1)\Gamma(s_2)}{\Gamma(s_1 + s_2)} = \int_0^{\pi/2} \frac{\cos^{s_1-1}\phi \sin^{s_2-1}\phi}{(\sin \phi + \cos \phi)^{s_1+s_2}}\, d\phi.$$

Finally, if spherical coordinates are used in (8): $z_1 = r \cos \phi \sin \theta$, $r_2 = r \sin \phi \sin \theta$, $z_3 = r \cos \theta$, we obtain

$$\frac{\Gamma(s_1)\Gamma(s_2)\Gamma(s_3)}{\Gamma(s_1 + s_2 + s_3)} = \int_0^{\pi/2} \int_0^{\pi/2} \frac{\cos^{s_1-1}\phi \, \sin^{s_2-1}\phi \, \sin^{s_1+s_2-1}\theta \, \cos^{s_3-1}\theta}{[(\sin\phi + \cos\phi)\sin\theta + \cos\theta]^{s_1+s_2+s_3}} \, d\phi \, d\theta$$

whence, in the special case of $s_1 = r$, $s_2 = 1 - r$, $s_3 = s$,

$$\int_0^{\pi/2} \int_0^{\pi/2} \frac{\cos^{r-1}\phi \, \sin^{-r}\phi \, \cos^{s-1}\theta}{[(\sin\phi + \cos\phi)\sin\theta + \cos\theta]^{s+1}} \, d\phi \, d\theta = \frac{\pi}{s \sin \pi r}.$$

For the infinitesimal generator G of $\{D_s\}$ we have formally by (5) and (6) the following integral representation

$$\begin{aligned}
Gf(x) &= \lim_{s \to 0} s^{-1}\left[\frac{1}{\Gamma(s)}\int_0^{\infty} z^{s-1}f(xe^{-z}) \, dz - f(x)\right] \\
&= \lim_{s \to 0}[s\Gamma(s)]^{-1}\left[\int_0^{\infty} z^{s-1}(f(xe^{-z}) - e^{-z}f(x)) \, dz\right] \\
&= \int_0^{\infty} \frac{f(xe^{-z}) - e^{-z}f(x)}{z} \, dz.
\end{aligned}$$

Since $D_s = e^{sG}$, by taking $f(x) = x/(1 + x)$, applying (6), and putting $x = 1$, we get the McLaurin series for the entire function $(1 - 2^{1-s})\zeta(s)$:

$$(1 - 2^{1-s})\zeta(s) = e^{sG}\left.\frac{x}{1 + x}\right|_{x=1} = \sum_{n=0}^{\infty} \frac{s^n}{n!} \left. G^n \frac{x}{1 + x}\right|_{x=1}.$$

By expressing the powers of G explicitly as multiple integrals, we have

$$(1 - 2^{1-s})\zeta(s) = a_0 + \frac{a_1 s}{1!} + \frac{a_1 s^2}{2!} + \cdots \tag{9}$$

where

$$a_0 = \frac{1}{2},$$

$$a_1 = \frac{1}{2}\int_0^{\infty} \frac{e^x - 1}{e^x + 1}\frac{dx}{xe^x} = \frac{1}{2}\log\frac{\pi}{2},$$

$$a_2 = \frac{1}{2}\int_0^{\infty}\int_0^{\infty} \frac{e^x - 1}{e^x + 1}\frac{e^y - 1}{e^y + 1}\frac{e^{x+y} - 1}{e^{x+y} + 1}\frac{dx}{xe^x}\frac{dy}{ye^y}. \tag{10}$$

Formulas (9) and (10) allow us to compute certain sums associated with the zeros of the zeta-function $\zeta(s)$. It is known [73] that $\zeta(s)$ has the factorization

$$\zeta(s) = \frac{e^{bs}}{2(s - 1)\Gamma(1 + s/2)} \prod_{\rho} [(1 - s/\rho)e^{s/\rho}] \tag{11}$$

where

$$b = -1 - \gamma/2 + \log 2\pi;$$

γ is the Euler constant, and the product is taken over all the so-called non-trivial zeros of $\zeta(s)$. These are complex numbers satisfying the relation $0 <$ Re $\rho < 1$; the still unproved Riemann hypothesis asserts that every ρ has real part $1/2$. The name "nontrivial zero" is due to the fact that by (11) $\zeta(s)$ also has zeros at all the poles of $\Gamma(1 + s/2)$, i.e., at $s = -2, -4, -6, \ldots$; these are called the trivial zeros of $\zeta(s)$.

We observe the power series due to Nielsen [55] for the entire function $1/\Gamma(1 + x)$:

$$\frac{1}{\Gamma(1 + x)} = 1 + \gamma x + \left(\frac{\gamma^2}{2} - \frac{\pi^2}{12}\right)x^2 + \cdots.$$

Now we use the above in (11) and then substitute the result for $\zeta(s)$ into (9). Expanding the factor $1 - 2^{1-s}$ in powers of s and equating the coefficients of like powers, we obtain after some formal manipulation

$$\sum_{\rho} \rho^{-2} = 1 + \log^2 2\pi - 4 \log 2 \log 2\pi + 2 \log^2 2 - \frac{\pi^2}{24} - 2a_2$$

where a_2 is given in (10). Similar procedure will enable us to compute the sums

$$\sum_{\rho} \rho^{-k}, \quad k \geq 2,$$

where ρ runs over all the nontrivial zeros of $\zeta(s)$, in terms of the coefficients $a_2, a_3, \ldots, a_k$.

10. GENERALIZATIONS OF TAYLOR'S SERIES

Taylor series may be written

$$f(z) = f(a) + \frac{f'(a)}{1!}(z - a) + \frac{f''(a)}{z!}(z - a)^2 + \cdots \tag{1}$$

and in this form it supplies an expansion of $f(z)$ in the powers of $z - a$. The special case $a = 0$ is known as McLaurin series. A generalization known as the Bürmann, Bürmann–Lagrange, or Bürmann–Lagrange–Taylor series supplies analogously the expansion of $f(z)$ in the powers of $\phi(z) - \phi(a)$ where $\phi(z)$ is an analytic function:

$$f(z) = f(a) + \sum_{n=1}^{\infty} \frac{1}{n!} \frac{d^{n-1}\left[f'(z)\dfrac{z - a}{\phi(z) - \phi(a)}\right]^n}{dz^{n-1}}\Bigg|_{z=a} [\phi(z) - \phi(a)]^n. \tag{2}$$

We notice that for $\phi(z) = z$, (2) becomes (1). In the special case when $a = \phi(a) = 0$ we have the Bürmann–Lagrange–McLaurin expansion

$$f(z) = f(0) + \sum_{n=1}^{\infty} \frac{1}{n!} \frac{d^{n-1} \left[f'(z) \dfrac{z}{\phi(z)} \right]^n}{dz^{n-1}} \Bigg|_{z=0} \phi^n(z). \tag{3}$$

The Lagrange form of (2) is somewhat different. Let $\phi(z)$ be analytic and let

$$z = a + x\phi(z). \tag{4}$$

By the implicit function theorem there is a unique root $z = z(x)$ which reduces to a at $x = 0$. If $f(z)$ is an analytic function then $f(z) = f(z(x))$ may be expanded in powers of x by the Lagrange formula

$$f(z) = f(a) + \sum_{n=1}^{\infty} \frac{d^{n-1} [\phi^n(a) f'(a)]}{da^{n-1}} \frac{x^n}{n!}. \tag{5}$$

Rewriting this as

$$f(z) = f(a) + \sum_{n=1}^{\infty} \frac{d^{n-1} [\phi^n(z) f'(z)]}{dz^{n-1}} \Bigg|_{z=a} \frac{x^n}{n!}$$

and replacing in (4) $\phi(z)$ by

$$\frac{z - a}{\phi(z) - \phi(a)},$$

we have formally (2) since now $x = \phi(z) - \phi(a)$. In particular, if $f(z) = z$ in (5) we have the expansion of the root $z(x)$ of (4):

$$z = a + \sum_{n=1}^{\infty} \frac{d^{n-1} \phi^n(a)}{da^{n-1}} \frac{x^n}{n!}. \tag{6}$$

(5) may be proved by a device due to Laplace [26]. Recalling that (4) determines z as a function of two variables, we write $z = z(x, a)$, $u = f(z)$. Differentiating (4) with respect to x and to a, we have

$$\frac{\partial z}{\partial a} = 1 + x\phi' \frac{\partial z}{\partial a}, \qquad \frac{\partial z}{\partial x} = \phi(z) + x\phi' \frac{\partial z}{\partial x}$$

so that

$$\frac{\partial u}{\partial x} = \frac{\partial u}{\partial a} \phi(z). \tag{7}$$

Let $F(z)$ be an arbitrary function, then

$$\frac{\partial \left[F(z) \dfrac{\partial u}{\partial x} \right]}{\partial a} = \frac{\partial \left[F(z) \dfrac{\partial u}{\partial a} \right]}{\partial x} \tag{8}$$

for both derivatives are equal to

$$F'(z)f'(z)\frac{\partial z}{\partial a}\frac{\partial z}{\partial x} + F(z)\frac{\partial^2 u}{\partial a\, \partial x}.$$

We prove by induction that for $n = 1, 2, \ldots$

$$\frac{\partial^n u}{\partial x^n} = \frac{\partial^{n-1}\left[\phi^n(z)\dfrac{\partial u}{\partial a}\right]}{\partial a^{n-1}}. \tag{9}$$

First, (9) holds for $n = 1$ by (7). Suppose next that it holds for all integers $\leq n$ and differentiate with respect to x:

$$\frac{\partial^{n+1} u}{\partial x^{n+1}} = \frac{\partial}{\partial x}\left[\frac{\partial^{n-1}\left[\phi^n(z)\dfrac{\partial u}{\partial a}\right]}{\partial a^{n-1}}\right] = \frac{\partial^{n-1}\dfrac{\partial\left[\phi^n(z)\dfrac{\partial u}{\partial a}\right]}{\partial x}}{\partial a^{n-1}}.$$

Applying (8) with $F(z) = \phi^n(z)$ we have

$$\frac{\partial^{n+1} u}{\partial x^{n+1}} = \frac{\partial^n\left[\phi^n(z)\dfrac{\partial u}{\partial x}\right]}{\partial a^n}$$

so that by (7)

$$\frac{\partial^{n+1} u}{\partial x^{n+1}} = \frac{\partial^n\left[\phi^{n+1}(z)\dfrac{\partial u}{\partial a}\right]}{\partial a^n}$$

which is (9) with $n + 1$ in place of n, thus concluding the course of induction.

At $x = 0$ we have $z = a$ and $\partial u/\partial a = f'(a)$ so that by (9)

$$\frac{\partial^n u}{\partial x^n}\bigg|_{x=0} = \frac{d^{n-1}[\phi^n(a)f'(a)]}{da^{n-1}}$$

and therefore (5) follows by applying the ordinary McLaurin expansion to $u = f(x)$ in powers of x.

EXAMPLE 1. Sum the series

$$S(c, w) = \sum_{n=0}^{\infty} \frac{(c + n)^n}{n!} w^n.$$

By the ratio-test the radius of convergence is $1/e$. We start by applying the Bürmann–McLaurin expansion (3) to the function $f(z) = e^{bz}$ in the powers of $\phi(z) = ze^{-z}$; we have then

$$\left(\frac{z}{\phi(z)}\right)^n = e^{nz}, \qquad f'(z) = be^{bz}$$

so that the coefficient of $\phi^n(z)$ is

$$\frac{d^{n-1}[be^{(b+n)z}]}{dz^{n-1}}\bigg|_{z=0} = b(b+n)^{n-1}$$

and hence

$$e^{bz} = 1 + \sum_{n=1}^{\infty} \frac{b(b+n)^{n-1}}{n!} w^n$$

where $w = \phi(z) = ze^{-z}$. Differentiating this with respect to w yields by the chain rule

$$\frac{e^{(b+1)z}}{1-z} = \sum_{n=0}^{\infty} \frac{(b+n+1)^n}{n!} w^n$$

so that putting $b + 1 = c$ we have

$$\sum_{n=0}^{\infty} \frac{(c+n)^n}{n!} w^n = \frac{e^{cz}}{1-z}. \qquad (10)$$

In particular, with $c = 0$ we have

$$\sum_{n=0}^{\infty} \frac{n^n}{n!} w^n = \frac{1}{1-z}. \qquad (11)$$

As shown in Figure 1

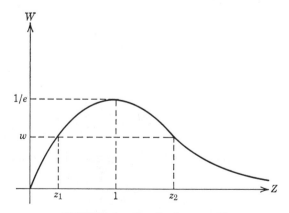

FIGURE 1.   Graph of $w = ze^{-z}$.

the equation $w = ze^{-z}$, for $0 < w < 1/e$, has exactly two roots z_1, z_2 ; by the continuity considerations we should have $z \to 0$ as $w \to 0$; hence in (10) and (11) z is the smaller if the two roots of $w = ze^{-z}$. More generally, the expansion of e^{bx} in the powers of $y = xe^{ax}$ is

$$e^{bx} = 1 + by + \frac{b(b - 2a)}{2!} y^2 + \frac{b(b - 3a)^2}{3!} y^3 + \frac{b(b - 4a)^3}{4!} y^4 + \cdots.$$

If we put $b = 1$, $a = -1$, $e^x = u$, we obtain the Eisenstein series

$$u = 1 + y + \frac{3y^2}{2!} + \frac{4^2 y^3}{3!} + \frac{5^3 y^4}{4!} + \cdots$$

for that root of the equation $\log u = yu$ $(y < 1/e)$ which reduces to 1 for $y = 0$.

EXAMPLE 2. Kepler's equation for the eccentric anomaly z arises in astronomy; it is

$$z = a + x \sin z.$$

By (6), the root which reduces to a at $x = 0$ is

$$z = a + \sum_{n=1}^{\infty} \frac{d^{n-1}[\sin^n a]}{da^{n-1}} \frac{x^n}{n!};$$

it was shown by Laplace that this converges for $|x| < .662743\ldots$, [26].

EXAMPLE 3. The equation for z

$$z = a + x \frac{z^2 - 1}{2} \tag{12}$$

has one root which is a when $x = 0$; by (6) this is given as

$$z = a + \sum_{n=1}^{\infty} \frac{d^{n-1}[(a^2 - 1)^n]}{da^{n-1}} \frac{x^n}{2^n n!};$$

differentiating with respect to a we have

$$\frac{\partial z}{\partial a} = \sum_{n=0}^{\infty} \frac{d^n[(a^2 - 1)^n]}{da^n} \frac{x^n}{2^n n!}. \tag{13}$$

On the other hand, (12) can be solved by the quadratic formula and the root we want is

$$z = \frac{1 - \sqrt{1 - 2ax + x^2}}{x}$$

so that

$$\frac{\partial z}{\partial a} = \frac{1}{\sqrt{1 - 2ax + x^2}}$$

which has the expansion in the series of Legendre polynomials

$$\frac{1}{\sqrt{1 - 2ax + x^2}} = \sum_{n=0}^{\infty} P_n(a)x^n.$$

Comparing this with (13) we have the Rodriguez formula for the nth Legendre polynomial:

$$P_n(a) = \frac{1}{2^n n!} \frac{d^n[(a^2 - 1)^n]}{da^n}.$$

Hence, expanding $(a^2 - 1)^n$ by the binomial theorem and differentiating n times with respect to a,

$$P_n(a) = 2^{-n} \sum_{k=0}^{n/2} (-1)^k \binom{n}{k} \binom{2n - 2k}{n} a^{n-2k}.$$

This may also be written as

$$P_n(a) = 2^{-n} \sum_{k=0}^{n/2} (-1)^k \binom{n-k}{k} \binom{2n-2k}{n-k} a^{n-2k}$$

or, in the form due to R. Murphy [62],

$$P_n(a) = \sum_{k=0}^{n} \binom{n+k}{n-k} \binom{2k}{k} \left(\frac{a-1}{2}\right)^k.$$

EXAMPLE 4. If

$$x^{n+1} + ax - b = 0$$

there is exactly one root $x = x(a, b)$ which reduces to 0 for $b = 0$. By Lagrange's theorem we find that for sufficiently small b

$$x(a, b) = \frac{b}{a} - \frac{b^{n+1}}{a^{n+2}} + \frac{2n + 2}{2!} \frac{b^{2n+1}}{a^{2n+3}} - \frac{(3n + 2)(3n + 3)}{3!} \frac{b^{3n+1}}{a^{3n+4}}$$

$$+ \frac{(4n + 2)(4n + 3)(4n + 4)}{4!} \frac{b^{4n+1}}{a^{4n+5}} - \cdots.$$

EXAMPLE 5. Let c_n be the number of labeled trees on n vertices. For $n = 2$ and 3 all such trees are shown in Figure 2a. For $n = 4$ we have $c_4 = 16$; there are 12 trees of the type of Figure 2b (12 is 4!/2, all 4! permutations but 1243 not counted as distinct from 3421, hence division by 2), and 4 trees of the

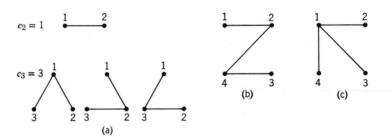

FIGURE 2.   Labeled trees.

type of Figure 2c. We let $t_n = nc_n$ be the number of rooted labeled trees (a rooted tree has one distinguished vertex, called the root) and put $t_1 = c_1 = 1$. If

$$T(x) = \sum_{n=1}^{\infty} t_n \, x^n / n!$$

is the exponential generating function for t_n's then it may be proved that $T(x) - xe^{T(x)} = 0$. Using Lagrange's formula (4), (5) with $a = 0$, $\phi(z) = e^z$, and T as z, we find

$$T(x) = \sum_{n=1}^{\infty} n^{n-1} \, x^n / n!$$

so that $t_n = n^{n-1}$, and we obtain the Cayley formula $c_n = n^{n-2}$.

11. BLISSARD–LUCAS CALCULUS AND THE BERNOULLI NUMBERS

In this section we continue with symbolic method. This is also known as the umbral or Blissard or Blissard–Lucas calculus; for its foundation, motivation, and many applications see [4], [61], and [62]. Let

$$f(x) = a_0 \, x^n + a_1 x^{n-1} + \cdots + a_{n-1} x + a_n$$

and let the sequence $b_1, b_2, \ldots, b_n$ satisfy

$$a_0 b_n + a_1 b_{n-1} + \cdots + a_{n-1} b_1 + a_n = 0. \tag{1}$$

In the Blissard–Lucas (BL) notation we write (1) symbolically as $f(b) = 0$. That is, after expanding $f(b)$ in powers of b we lower the index by replacing b^k with b_k. Let $b_0 = 1$ and define the sequence $b_1, b_2, \ldots$ by

$$2b_1 + 1 = 0, \qquad 3b_2 + 3b_1 + 1 = 0, \qquad 4b_3 + 6b_2 + 4b_1 + 1 = 0, \ldots$$

which in the BL notation may be written as

$$(b + 1)^n - b^n = 0, \qquad n = 2, 3, \ldots. \tag{2}$$

Solving successively for the b's we have $b_1 = -1/2, b_2 = 1/6, b_4 = -1/30, \ldots,$ $b_3 = b_5 = \cdots = 0$. If $f(x)$ is a polynomial then by Taylor's theorem

$$f(x + b + 1) - f(x + b)$$

$$= f'(x) + [(b + 1)^2 - b^2]\frac{f'(x)}{2!} + [(b + 1)^3 - b^3]\frac{f''(x)}{3!} + \cdots$$

so that in the BL-symbolism

$$f(x + b + 1) - f(x + b) = f'(x). \tag{3}$$

For the special case of $f(x) = x^{n+1}/(n + 1)$ we have $f'(x) = x^n$; hence, putting in (3) x successively equal to $0, 1, \ldots, N - 1$ and applying telescoping cancellation, we have

$$\sum_{j=1}^{N-1} j^n = \frac{(N + b)^{n+1} - b^{n+1}}{n + 1}.$$

If $S_n(N - 1)$ denotes the left-hand side, as in Section 1, then developing the right-hand side by the BL-method we find

$$S_n(N - 1) = \sum_{k=0}^{n} \frac{1}{n + 1}\binom{n + 1}{k}b_k N^{n+1-k}.$$

Let us determine the exponential generating function of the sequence $b_0, b_1, b_2, \ldots$:

$$f(x) = \sum_{k=0}^{\infty} \frac{b_k}{k!} x^k.$$

In the BL-terminology this is simply e^{bx}; hence by (2)

$$e^{(b+1)x} - e^{bx} = \sum_{k=0}^{\infty} \frac{(b + 1)^k - b^k}{k!} x^k = x$$

since every coefficient vanishes except for $k = 1$. Therefore

$$e^{(b+1)x} - e^{bx} = e^{bx}(e^x - 1) = x$$

so that

$$e^{bx} = \frac{x}{e^x - 1},$$

$$\sum_{k=0}^{\infty} \frac{b_k}{k!} x^k = \frac{x}{e^x - 1}. \tag{4}$$

The coefficient b_k is called the kth Bernoulli number although sometimes it is also

$$B_k = (-1)^{k-1} b_{2k}, \qquad k = 1, 2, \ldots$$

which is called the kth Bernoulli number. We shall use both the numbers b_k and B_k.

Along with the Bernoulli numbers we have the Bernoulli polynomials $b_n(x)$ defined by the symbolic BL-relation.

$$b_n(x) = (b + x)^n. \qquad (5)$$

We have therefore

$$b_n(x) = \sum_{k=0}^{n} \binom{n}{k} b_k x^{n-k}; \qquad (6)$$

differentiating (5) with respect to x

$$\frac{db_n(x)}{dx} = n(b + x)^{n-1} = n b_{n-1}(x)$$

and generally

$$\frac{d^k b_n(x)}{dx^k} = \frac{n!}{(n-k)!} b_{n-k}(x); \qquad (7)$$

using the binomial theorem twice

$$b_n(x + 1) - b_n(x) = (x + b + 1)^n - (x + b)^n = \sum_{k=0}^{n} \binom{n}{k} x^k [(b+1)^{n-k} - b^{n-k}];$$

we observe that by (2) every square bracket vanishes except when $n - k = 1$, so that

$$b_n(x + 1) - b_n(x) = n x^{n-1}; \qquad (8)$$

finally, we have the generating function

$$\sum_{n=0}^{\infty} \frac{b_n(x)}{n!} z^n = \sum_{n=0}^{\infty} \frac{(b+x)^n z^n}{n!} = e^{(b+x)z} = e^{xz} e^{bz}$$

and since e^{bz} is $z/(e^z - 1)$ by (4),

$$\sum_{n=0}^{\infty} \frac{b_n(x)}{n!} z^n = \frac{z e^{xz}}{e^z - 1}. \qquad (9)$$

Expanding $b_n(x)$ in a Fourier series:

$$b_n(x) = \frac{a_0}{2} + \sum_{m=1}^{\infty} a_m \cos 2m\pi x + \sum_{m=1}^{\infty} b_m \sin 2m\pi x, \qquad 0 \le x < 1;$$

we compute the coefficients by integration by parts and the previously developed formulas and we find that

$$b_{2n}(x) = 2(2n)!(-1)^{n-1} \sum_{m=1}^{\infty} \frac{\cos 2m\pi x}{(2\pi m)^{2n}},$$

$$b_{2n-1}(x) = 2(2n-1)!(-1)^n \sum_{m=1}^{\infty} \frac{\sin 2m\pi x}{(2\pi m)^{2n-1}}.$$

Hence, putting $x = 0$, we find that

$$\sum_{m=1}^{\infty} \frac{1}{m^{2n}} = \frac{(2\pi)^{2n}}{2(2n)!} B_n. \tag{10}$$

Our McLaurin series for $\sec x$ from Section 8,

$$\sec x = A_0 + \frac{A_2}{2!} x^2 + \frac{A_4}{4!} x^4 + \cdots,$$

could be similarly handled by the BL-method. We put $E_0 = 1$ and we introduce the sequence $E_0, E_1, E_2, \ldots$ by the defining BL-relation

$$(E + 1)^n + (E - 1)^n = 0, \qquad n = 1, 2, \ldots \tag{11}$$

which gives us $E_1 = E_3 = \cdots = 0$, $E_0 = 1$, $E_2 = -1$, $E_4 = 5$, etc. If

$$e^{Ex} = \sum_{n=0}^{\infty} \frac{E_n x^n}{n!}$$

is the exponential generating function, then by (11)

$$e^{(E+1)x} + e^{(E-1)x} = \sum_{n=0}^{\infty} \frac{(E+1)^n + (E-1)^n}{n!} x^n$$

is 2 because every term except for $n = 0$ vanishes. Hence

$$e^{Ex}(e^x + e^{-x}) = 2$$

or

$$e^{Ex} = \operatorname{sech} x, \qquad e^{iEx} = \sec x.$$

Comparison with Section 8 shows that $A_{2k} = (-1)^k E_{2k}$. The numbers E_{2n} are called after their discoverer the Euler numbers and from their genesis the secant coefficients.

12. POISSON SUMMATION FORMULA

For $0 \leq x < 2\pi$ a function $F(x)$ continuous together with its derivative may be represented by the complex Fourier series

$$F(x) = \frac{1}{2}\pi \sum_{k=-\infty}^{\infty} e^{ikx} \int_0^{2\pi} e^{-iku}F(u)\,du.$$

Suppose that

$$F(x) = \sum_{n=-\infty}^{\infty} f(x + 2n\pi)$$

where the series and its derivatives converge sufficiently well to justify the necessary operations (uniform and absolute convergence of the series and its derivative will do). Substituting the series for F we have

$$\sum_{n=-\infty}^{\infty} f(x + 2n\pi) = \frac{1}{2\pi} \sum_{k=-\infty}^{\infty} e^{ikx} \sum_{n=-\infty}^{\infty} \int_0^{2\pi} e^{-iku}f(u + 2\pi)\,du.$$

Since the integrand is periodic with the period 2π we may integrate over any interval of length 2π, hence

$$\sum_{n=-\infty}^{\infty} \int_0^{2\pi} = \sum_{n=-\infty}^{\infty} \int_{2n\pi}^{2(n+1)\pi} = \int_{-\infty}^{\infty}$$

so that

$$\sum_{n=-\infty}^{\infty} f(x + 2n\pi) = \frac{1}{2\pi} \sum_{k=-\infty}^{\infty} e^{ikx} \int_{-\infty}^{\infty} e^{-iku}f(u)\,du.$$

Putting $x = 0$ and introducing $g(x) = f(2\pi x)$ we obtain

$$\sum_{n=-\infty}^{\infty} g(n) = \sum_{n=-\infty}^{\infty} \int_{-\infty}^{\infty} e^{-2\pi i n u}g(u)\,du \qquad (1)$$

or, if g is real,

$$\sum_{n=-\infty}^{\infty} g(n) = \sum_{n=-\infty}^{\infty} \int_{-\infty}^{\infty} g(u)\cos 2\pi n u\,du; \qquad (2)$$

either (1) or (2) is called the Poisson summation formula.

EXAMPLE 1. Let

$$\theta(x) = \sum_{n=-\infty}^{\infty} e^{-n^2\pi x}, \qquad x > 0,$$

be the elliptic theta function; then we have the functional equation

$$\theta(x) = \frac{1}{\sqrt{x}}\,\theta\left(\frac{1}{x}\right). \qquad (3)$$

To prove it we apply (1), or, preferably, (2) to the function $g(x) = e^{-\pi u^2 x}$. Similarly, we obtain the more general formula

$$\sum_{n=-\infty}^{\infty} e^{-n^2\pi x + 2n\pi ax} = \frac{1}{\sqrt{x}} e^{\pi a^2 x}\left[1 + 2\sum_{n=1}^{\infty} e^{-n^2\pi/x}\cos 2\pi na\right],$$

which may also be written

$$\sum_{n=-\infty}^{\infty} e^{-n^2\pi x + 2n\pi ax} = \frac{1}{\sqrt{x}} e^{\pi a^2 x} \sum_{n=-\infty}^{\infty} e^{2\pi ina - \pi n^2/x}. \tag{4}$$

We give now an independent proof of (4) due to Polya. The spirit of this proof is somewhat similar to that of Darboux's method of factorizing $\sin x$, given in Section 6. Also, we use the arithmetic "thinning out" of Section 5, and we observe that the final passage to the limit parallels the derivation of the normal distribution from the binomial one.

By the binomial theorem

$$(z^{1/2} + z^{-1/2})^{2m} = \sum_{k=-m}^{m} \binom{2m}{m+k} z^k.$$

Let p be an even integer and let $u = e^{2\pi i/p}$; we replace in the above z by zu^j and we sum over j from $-p/2$ to $p/2$:

$$\sum_{j=-p/2}^{p/2} [(zu^j)^{1/2} + (zu^j)^{-1/2}]^{2m} = \sum_{j=-p/2}^{p/2} \sum_{k=-m}^{m} \binom{2m}{m+k} z^k u^{kj}.$$

Since

$$\sum_{j=-p/2}^{p/2} u^{kj} = p \quad \text{if } k \mid p$$
$$= 0 \quad \text{if } k \nmid p$$

we can write the above as

$$\sum_{j=-p/2}^{p/2} [(zu^j)^{1/2} + (zu^j)^{-1/2}]^{2m} = \sum_{-\frac{m}{p} \le k \le \frac{m}{p}} p\binom{2m}{m+kp} z^{kp}.$$

Dividing both sides by 2^{2m} and writing $z = e^{s/p}$ we have

$$\sum_{j=-p/2}^{p/2} \left[1 + \frac{(s+2\pi ij)^2}{8p^2} + 0\left(\frac{1}{p^3}\right)\right]^{2m} = \sum_{-\frac{m}{p} \le k \le \frac{m}{p}} 2^{-2m} p\binom{2m}{m+kp} e^{sk}. \tag{5}$$

We now justify the passage to the limit when

$$p \to \infty, \quad m \to \infty, \quad \frac{m}{p^2} \to t$$

with t finite. Observing the exponential limit

$$\lim_{n \to \infty} \left[1 + \frac{x}{n} + 0\left(\frac{1}{n^{1+\alpha}}\right) \right]^n = e^x, \qquad \alpha > 0,$$

and the Stirling formula for the factorial, which gives us

$$\lim_{m, p \to \infty} 2^{-2m} p\binom{2m}{m+kp} = \frac{1}{\sqrt{\pi t}} e^{-k^2/t},$$

we get from (5)

$$\sum_{j=-\infty}^{\infty} e^{t(s+2\pi i j)^2/4} = \frac{1}{\sqrt{\pi t}} \sum_{k=-\infty}^{\infty} e^{sk-k^2/t};$$

changing the variables from s and t to x and a: $t = 1/\pi x$ and $s = 2\pi ax$, we get (4).

EXAMPLE 2. Let

$$\psi(x) = \sum_{n=1}^{\infty} e^{-n^2 \pi x}$$

so that $\theta(x) = 1 + 2\psi(x)$; (3) becomes then

$$\psi(x) = \frac{1}{\sqrt{x}} \psi\left(\frac{1}{x}\right) + \frac{1}{2\sqrt{x}} - \frac{1}{2}. \qquad (6)$$

We may deduce from (6) the functional equation for the Riemann zeta-function $\zeta(s)$. For, we have

$$\int_0^\infty x^{(s/2)-1} e^{-n^2 \pi x} \, dx = \frac{\Gamma\left(\dfrac{s}{2}\right)}{\pi^{s/2} n^s}$$

whence

$$\pi^{-(s/2)} \Gamma(s/2) \zeta(s) = \int_0^\infty x^{(s/2)-1} \psi(x) \, dx;$$

we split the integration:

$$\int_0^\infty = \int_0^1 + \int_1^\infty$$

and apply (4) to the finite integral; when this is done we take $1/x$ as the new variable, obtaining

$$\pi^{-(s/2)} \Gamma\left(\frac{s}{2}\right) \zeta(s) + \frac{1}{s(1-s)} = \int_1^\infty [x^{(s/2)-1} + x^{-(1+s)/2}] \psi(x) \, dx.$$

Since the right-hand side is invariant when s is replaced by $1 - s$, so is the left-hand side:

$$\pi^{-s/2}\Gamma\left(\frac{s}{2}\right)\zeta(s) = \pi^{-(1-s)/2}\Gamma\left(\frac{1-s}{2}\right)\zeta(1-s)$$

which is the symmetric form of the functional equation for $\zeta(s)$.

Some deep results in number theory have been obtained by generalizing the methods of this section under the following scheme:

real line $(-\infty, \infty)$ $\quad \to$ a topological group G,

integers $\ldots, -1, 0, 1, \ldots \to$ dual group of G,

exponentials $e^{-2\pi n i u}$ $\quad \to$ group characters,

Fourier series $\quad\quad\quad \to$ Peter–Weyl expansions.

13. DARBOUX AND EULER–McLAURIN FORMULAS

Let $\phi(t)$ be a polynomial of degree n and let $f(x)$ be sufficiently differentiable. Then by differentiating and applying the telescoping cancellation we verify that

$$\frac{d}{dt} \sum_{k=1}^{n} (-1)^k (x-a)^k \phi^{(n-k)}(t) f^{(k)}[a + t(x-a)]$$

$$= -(x-a)\phi^{(n)}(t)f'[a + t(x-a)]$$

$$+ (-1)^n (x-a)^{n+1} \phi(t) f^{(n+1)}[a + t(x-a)].$$

Hence, by integrating with respect to t from 0 to 1, we obtain the Darboux formula

$$\phi^{(n)}(0)[f(x) - f(a)] = \sum_{k=1}^{n} (-1)^{k-1}(x-a)^k [\phi^{(n-k)}(1)f^{(k)}(x) - \phi^{(n-k)}(0)f^{(k)}(a)]$$

$$+ (-1)^n (x-a)^{n+1} \int_0^1 \phi(t) f^{(n+1)}[a + t(x-a)]\, dt. \quad (1)$$

Various important theorems are special cases of (1). For instance, choosing $\phi(t) = (t-1)^n$, we obtain the Taylor theorem with remainder:

$$f(x) = \sum_{k=0}^{n} \frac{f^{(k)}(a)}{k!}(x-a)^k + (-1)^n \frac{(x-a)^{n+1}}{n!}$$

$$\times \int_0^1 (t-1)^n f^{(n+1)}[a + t(x-a)]\, dt.$$

Replacing in (1) n by $2n$ and choosing $\phi(t)$ to be the Bernoulli polynomial $b_{2n}(t)$, we obtain, after some evaluations based on equations (5)–(8) of Section 11, the Euler–McLaurin formula with the remainder term:

$$(x - a)f'(a) = f(x) - f(a) - \frac{x - a}{2}[f'(x) - f'(a)]$$

$$+ \sum_{k=1}^{n-1} \frac{(-1)^{k-1}B_k}{(2k)!}(x - a)^{2k}[f^{(2k)}(x) - f^{(2k)}(a)]$$

$$- \frac{(x - a)^{2n+1}}{(2n)!}\int_0^1 b_{2n}(t)f^{(2n+1)}[a + (x - a)t]\,dt.$$

The more usual form of this is obtained by writing $F(x)$ for $f'(x)$ and $a + h$ for x:

$$\int_a^{a+h} F(x)\,dx = \frac{h}{2}[F(a) + F(a + h)]$$

$$+ \sum_{k=1}^{n-1} \frac{(-1)^k B_k h^{2k}}{(2k)!}[F^{(2k-1)}(a + h) - F^{(2k-1)}(a)]$$

$$+ \frac{h^{2n+1}}{(2n)!}\int_0^1 b_{2n}(t)F^{(2n)}(a + th)\,dt. \qquad (2)$$

Writing $a, a + h, \ldots, a + (N - 1)h$ for a in (2), adding the results, and applying telescoping cancellation, we have

$$\int_a^{a+Nh} F(x)\,dx - h\sum_{k=0}^{N} F(a + kh)$$

$$= -\frac{1}{2}[F(a) + F(a + Nh)]$$

$$+ \sum_{k=1}^{N-1} \frac{(-1)^k B_k h^{2k}}{(2k)!}[F^{(2k-1)}(a + Nh) - F^{(2k-1)}(a)]$$

$$+ \frac{h^{2n+1}}{(2n)!}\int_0^1 b_{2n}(t)\left[\sum_{k=0}^{N-1} F^{(2n)}(a + kh + ht)\right]dt. \qquad (3)$$

This displays the Euler–McLaurin formula in both its roles: as a means of comparing an integral with an approximating finite sum, and as a means of expressing an interval-dependent functional, namely the left-hand side of (3), in terms of the differential conditions at the end-points only.

We observe that the Euler–McLaurin formula without the remainder term may be very quickly obtained by the symbolic method. If Δ_h is the difference operator

$$\Delta_h f(x) = f(x + h) - f(x)$$

then the inverse Δ_h^{-1} turns out to be the summation operator

$$\Delta_h^{-1}f(x) = \sum_j f(x + jh);$$

Taylor's theorem may be written symbolically as

$$\Delta_h = e^{hD} - 1$$

and so

$$\Delta_h^{-1} = \frac{1}{hD} \frac{hD}{e^{hD} - 1}.$$

Expanding in series with Bernoulli numbers for coefficients, we obtain the Euler–McLaurin formula.

EXAMPLE. Take in (3) $F(x) = \log x$, $a = h = 1$, $N = n - 1$. After some limit evaluations we obtain the Stirling formula

$$\log \frac{n!}{\sqrt{2\pi n}(n/e)^n} \sim \sum_{k=1}^{\infty} \frac{(-1)^{k+1}B_{2k+1}}{(2k+1)(2k+2)} n^{-1-2k}.$$

By formula (10) of Section 11 we have

$$B_n = 0[(2n)!(2\pi)^{-2n}], \qquad n \text{ large},$$

and so the preceding series diverges. It is, in fact, a so-called asymptotic series; for definitions and further analysis see Section 7 of Chapter 6. We observe here only that by taking exponentials we get

$$n! \sim (n/e)^n \sqrt{2\pi n} \left[1 + \frac{1}{12n} + \frac{1}{288n^2} - \frac{129}{51840n^3} + \cdots \right]$$

and this expression, although divergent, allows us to compute $n!$ for large n with considerable accuracy.

14. LAMBERT SERIES

A series of the type

$$\sum_{n=1}^{\infty} \frac{a_n x^n}{1 - x^n}$$

is called a Lambert series. The simplest case is

$$h(x) = \sum_{n=1}^{\infty} \frac{x^n}{1 - x^n} = \sum_{n=1}^{\infty} \sum_{m=1}^{\infty} x^{nm};$$

collecting together like powers we have

$$h(x) = \sum_{k=1}^{\infty} d(k)x^k \tag{1}$$

where $d(k)$ is the number of ways of expressing k as mn, i.e., the number of divisors of k with 1 and k themselves counted in. More generally

$$\sum_{n=1}^{\infty} \frac{g(n)x^n}{1-x^n} = \sum_{k=1}^{\infty} d_g(k)x^k$$

where

$$d_g(k) = \sum_{d|k} g(d) \tag{2}$$

the summation extending over all the divisors of k, 1 and k included. We find for instance that values $g(n) = 1$ or -1 can be found for $n = 1, 2, 3, \ldots$, so that

$$\sum_{n=1}^{\infty} \frac{g(n)x^n}{1-x^n} = \sum_{k=1}^{\infty} x^{k^2};$$

what are those values $g(n)$? One way to answer this is to make use of the Dirichlet operator D_s introduced in Section 9. We observe that if

$$\sum_{n=1}^{\infty} a_n x^n = f(x) \qquad \text{and} \qquad \sum_{n=1}^{\infty} \frac{a_n x^n}{1-x^n} = F(x) \tag{3}$$

then

$$D_s F(x)\bigg|_{x=1} = \zeta(s) D_s f(x)\bigg|_{x=1}. \tag{4}$$

Here we have

$$\sum_{n=1}^{\infty} g(n)x^n = f(x), \qquad \sum_{n=1}^{\infty} x^{n^2} = F(x)$$

so that

$$D_s F(x)\bigg|_{x=1} = \sum_{n=1}^{\infty} (n^2)^{-s} = \zeta(2s).$$

Therefore by (3) and (4)

$$D_s f(x)\bigg|_{x=1} = \sum_{n=1}^{\infty} \frac{g(n)}{n^s} = \frac{\zeta(2s)}{\zeta(s)}. \tag{5}$$

Now, it is known from the theory of generating functions of Dirichlet series [28] that

$$\frac{\zeta(2s)}{\zeta(s)} = \sum_{n=1}^{\infty} \frac{(-1)^{b(n)}}{n^s} \tag{6}$$

where $b(1) = 0$, and for $n > 1$, $b(n)$ is the total number of prime factors of n, counted with multiplicity; e.g., $b(4) = 2$, $b(8) = 3$, $b(12) = 3$, etc. By the identity theorem for Dirichlet series ($=$identical Dirichlet series have corresponding coefficients equal) we find from (5) and (6) that $g(n) = (-1)^{b(n)}$, so that

$$\sum_{n=1}^{\infty} \frac{(-1)^{b(n)} x^n}{1 - x^n} = \sum_{k=1}^{\infty} x^{k^2}.$$

By (2) this gives us

$$\sum_{d|k} (-1)^{b(d)} = 1 \qquad \text{if } k \text{ is a square}$$

$$= 0 \qquad \text{otherwise.}$$

Returning to (1), we observe that for $k > 1$ the coefficient of x^k in

$$h(x) = \sum_{n=1}^{\infty} \frac{x^n}{1 - x^n} = \sum_{k=1}^{\infty} d(k) x^k$$

is ≥ 2 and it is 2 if and only if k is prime. It is therefore easy to understand how this might have given mathematicians hope of being able to handle the difficult questions about primes via Lambert series. However, this turned out to be a prime example of a mathematical flash in the pan and nothing about primes was learned in that way. It was only when Riemann introduced his zeta function

$$\zeta(s) = \sum_{n=1}^{\infty} n^{-s} = \prod_{p} (1 - p^{-s})^{-1},$$

with p ranging over primes 2, 3, 5, ..., and started investigating its behavior for complex s, that progress began to be made. Let

$$g(x) = \sum_{n=1}^{\infty} a_n x^n, \qquad |x| < r,$$

and put

$$F(x) = \sum_{n=1}^{\infty} \frac{a_n x^n}{1 - x^n}, \qquad G(x) = \sum_{n=1}^{\infty} \frac{a_n x^n}{1 - x^{2n}}, \qquad H(x) = \sum_{n=1}^{\infty} \frac{a_n x^n}{1 + x^{2n}}.$$

Then for $|x| < \min(1, r)$ we have

$$F(x) = \sum_{m=1}^{\infty} g(x^m), \tag{7}$$

$$G(x) = \sum_{m=0}^{\infty} g(x^{2m+1}), \tag{8}$$

$$H(x) = \sum_{m=0}^{\infty} (-1)^m g(x^{2m+1}). \tag{9}$$

By taking suitable coefficients a_n we get a variety of identities. For instance, with $a_n = 1/n$ and $a_n = a^n$ we get from (7)

$$\sum_{n=1}^{\infty} \frac{1}{n} \frac{x^n}{1 - x^n} = \log \prod_{m=1}^{\infty} (1 - x^m)^{-1}$$

and

$$\sum_{n=1}^{\infty} \frac{a^n x^n}{1 - x^n} = \sum_{m=1}^{\infty} \frac{a x^m}{1 - a x^m},$$

and with $a_n = 1$ we find from (9) that

$$\sum_{n=1}^{\infty} \frac{x^n}{1 + x^{2n}} = \sum_{m=0}^{\infty} (-1)^m \frac{x^{2m+1}}{1 - x^{2m+1}} = h(x),$$

say. Since the left-hand side is unchanged on replacing x by $1/x$, we find that $h(x)$ represents one analytic function inside the unit circle C, another analytic function outside C, and C itself is a natural boundary for h. This is the first example of such a function; it is due to Weierstrass.

4

MISCELLANEOUS ELEMENTARY TOPICS

1. CHOOSING A SUITABLE ADDITIONAL UNKNOWN

To evaluate $x = \cos 20° \cos 40° \cos 80°$ fastest we introduce the unknown $y = \sin 20° \sin 40° \sin 80°$ and we have

$$8xy = (2 \sin 20° \cos 20°)(2 \sin 40° \cos 40°)(2 \sin 80° \cos 80°)$$
$$= \sin 40° \sin 80° \sin 160° = y$$

whence, since $y \neq 0$, $x = 1/8$. The reader may wish to find other sets of angles with similar recurrence under duplication. In particular, for two instead of three angles we have

$$C = \cos \alpha \cos \beta, \qquad S = \sin \alpha \sin \beta, \qquad 4SC = \sin 2\alpha \sin 2\beta$$

and for recurrence we need $2\alpha = \beta$, $2\beta = 180° - \alpha$, so that $\alpha = 36°$ while $\beta = 72°$. If we put $x = \cos 36°$ we find that

$$C = x(2x^2 - 1) = 1/4; \qquad 8x^3 - 4x - 1 = (2x + 1)(4x^2 - 2x - 1) = 0,$$

and solving this we have $\cos 36° = (1 + 5^{1/2})/4$.

Suppose that we tried to evaluate $x = \cos 20°$ from our first example; by the duplication formula for the cosine we have

$$x = \cos 20°, \qquad 2x^2 - 1 = \cos 40°, \qquad 2(2x^2 - 1)^2 - 1 = \cos 80°$$

so that

$$8x(2x^2 - 1)(8x^4 - 8x^2 + 1) - 1 = 0$$

or, multiplying out,

$$128x^7 - 192x^5 + 80x^3 - 8x - 1 = 0. \tag{1}$$

131

By the theorem of Gauss the rational roots $x = p/q$ (in reduced form) must satisfy the conditions $q \mid \pm 128$, $p \mid \pm 1$ which shows that $x = 1/2$ is the only rational root. Hence $2x - 1$ is a linear factor and by division in (1) we find the sextic

$$64x^6 + 32x^5 - 80x^4 - 40x^3 + 20x^2 + 10x - 1 = 0.$$

Is there any hope to achieve a complete factorization and so to solve for x by solving solely *linear* and *quadratic* equations? No, for otherwise $x = \cos 20°$ would be expressible by square roots alone and this would imply that the regular nonagon, of side $2r(1 - x^2)^{1/2}$, could be inscribed into a circle of radius r using ruler and compasses only. However, this is known to be impossible. On the other hand, $\cos 36°$ is expressible by square roots alone and the regular pentagon can be constructed by ruler and compasses.

2. MULTIPLICATIONS, ADDITIONS, AND COMPLEXITY

Suppose that we have to evaluate the four products

$$13 \cdot 29(= 377), \qquad 13 \cdot 34(= 442), \qquad 27 \cdot 29(= 783), \qquad 27 \cdot 34(= 918).$$

Suppose further that we have access to a multiplying machine which will multiply any two numbers, no matter now long, but that we are allowed only a single multiplication. Can all four products be evaluated? Yes, since

	1	300	000	027
			290	034
377	044		207	830	918

so that we get all four products separated by 0's. The reader may wish to generalize this: let $a_1, a_2, \ldots, a_n$ be n positive integers written in the common decimal form; let $b_1, b_2, \ldots, b_m$ be m other positive integers similarly presented; then there is a way of stringing together the two sequences with suitable spacings of *sufficiently many* 0's

$$x = a_1 00 \quad \ldots \quad 0a_2 00 \quad \ldots \quad 0a_n$$

$$y = b_1 00 \quad \ldots \quad 0b_2 00 \quad \ldots \quad 0b_m$$

so that all the mn products of the form $a_i b_j$ are obtained as distinct contiguous groups in xy, separated by 0's. This may be generalized to mnp products $a_i b_j c_k$ if we are allowed two multiplications, etc. The reader may wish to estimate the minimal lengths of x and y above, given n and m, and given the information that the numbers a_i and b_j need at most k digits each.

Continuing with similar material, we find that raising 1001 to successive powers results in the following values:

$$
\begin{aligned}
1001^1 &= & & 1\ 001 \\
1001^2 &= & & 1\ 002\ 001 \\
1001^3 &= & & 1\ 003\ 003\ 001 \\
1001^4 &= & & 1\ 004\ 006\ 004\ 001 \\
1001^5 &= & & 1\ 005\ 010\ 010\ 005\ 001 \\
1001^6 &= & & 1\ 006\ 015\ 020\ 015\ 006\ 001 \\
1001^7 &= & 1\ &007\ 021\ 035\ 035\ 021\ 007\ 001;
\end{aligned}
$$

these give us the binomial coefficients. If we raise to a few more powers there is a spill-over. The reader may wish to show similarly that if a and b are positive integers written in the decimal form, if n is any preassigned integer, and if in the expression

$$x = a00 \quad \ldots \quad 0b$$

there are sufficiently many 0's, then raising x to any power k, $1 \leq k \leq n$, we obtain all the terms of the binomial expansion

$$a^k, \binom{k}{1}a^{k-1}b, \binom{k}{2}a^{k-2}b^2, \ldots, \binom{k}{k-1}ab^{k-1}, b^k$$

as separate groups of digits spaced by 0's. Again, the reader may wish to estimate the necessary spacing of a and b in x.

Consider two 2-by-2 matrices with real numbers as entries, which are to be multiplied:

$$\begin{pmatrix} a & b \\ c & d \end{pmatrix}\begin{pmatrix} A & B \\ C & D \end{pmatrix} = \begin{pmatrix} m & n \\ p & q \end{pmatrix}.$$

We know that $m = aA + bC$, $n = aB + bD$, $p = cA + dC$, and $q = cB + dD$. It looks therefore that eight multiplications are necessary. However, at the cost of an increased number of additions we can multiply the two matrices using seven multiplications only. For if

$$
\begin{aligned}
x &= (a + d)(A + D) & v &= (c + d)A \\
y &= a(B - D) & w &= (c - a)(A + B) \\
z &= (a + b)D & t &= (b - d)(C + D), \\
u &= d(C - A)
\end{aligned}
$$

then it may be verified that

$$m = x + t - z + u, \quad n = y + z, \quad p = u + v, \quad q = x + w + y - v.$$

It has been proved by Strassen [70], Hopcroft and Kerr [33], and Winograd [78] that the number of multiplications cannot be reduced below seven. No such minimal number appears to be known for the product of two 3-by-3 matrices or in any higher case. Similarly, to multiply two complex numbers $a + bi$ and $c + di$ three multiplications suffice since

$$(a + bi)(c + di) = ac - bd + [(a + b)(c + d) - ac - bd]i,$$

and it is shown in [79] that at last three multiplications are necessary. Similar problems on the minimal number of multiplications may be raised for scalar products, polynomials, quadratic and higher forms, etc. For instance, the quadratic form

$$F = \sum_{1 \le j \le k < n} x_i x_j$$

requires at most $n - 1$ multiplications since

$$F = \sum_{i=1}^{n-1} \left(x_i \sum_{j=i+1}^{n} x_j \right);$$

it may be proved that no smaller number of multiplications will do. Here it will be observed that if F were evaluable by $n - k$ multiplications, then, since

$$G = \sum_{i=1}^{n} x_i^2 = \left(\sum_{i=1}^{n} x_i \right)^2 - F - F,$$

G would be evaluable by $n - k + 1$ multiplications. For the scalar product xy of two n-dimensional vectors Winograd [79] has shown that the standard component-by-component method of computation minimizes simultaneously the number of multiplications, n, and the number of additions, $n - 1$.

Of the many methods to compute $2^{1/2}$ to arbitrary accuracy, the following one—though not the most practical—is probably the simplest. We start with any positive integers p_0 and q_0 and we define

$$p_{n+1} = p_n + 2q_n, \qquad q_{n+1} = p_n + q_n, \qquad n = 0, 1, 2, \ldots;$$

then $p_n/q_n \to 2^{1/2}$. For it may be verified that

$$p_{n+1}^2 - 2q_{n+1}^2 = -(p_n^2 - 2q_n^2) = \cdots = (-1)^{n+1}(p_0^2 - 2q_0^2)$$

and hence

$$\frac{p_{n+1}}{q_{n+1}} - 2^{1/2} = \frac{(-1)^{n+1}(p_0^2 - 2q_0^2)}{q_{n+1}(p_{n+1} + 2^{1/2}q_{n+1})};$$

therefore the sequence p_0/q_0, p_1/q_1, ... converges to $2^{1/2}$ steadily and the successive terms are alternately below and above $2^{1/2}$. Thus

$$\left| \frac{p_n}{q_n} - 2^{1/2} \right| < \frac{1}{2} \left| \frac{p_n}{q_n} - \frac{p_{n-1}}{q_{n-1}} \right|$$

which gives us a convenient stopping rule when desired accuracy has been achieved.

We observe that the successive approximations p_n/q_n are generated iteratively by means of two additions of integers per iteration stage and no other operations (we need one addition for q_{n+1}: $q_{n+1} = p_n + q_n$, and another one for p_{n+1}: $p_{n+1} = q_{n+1} + q_n$). We ask now: What other numbers can be similarly computed? To pose it rigorously we confine ourselves to real irrational numbers; we start with suitable integers p_{10}, p_{20}, ..., p_{k0}, and we require $p_{1\,n+1}$, $p_{2\,n+1}$, ..., $p_{k\,n+1}$ to be generable from p_{1n}, p_{2n}, ..., p_{kn} by fixed additions:

$$p_{i\,n+1} = \sum_{j=1}^{k} c_{ij} p_{jn}, \qquad i = 1, 2, \ldots, k; \quad n = 0, 1, 2, \ldots \tag{1}$$

where the c_{ij} are fixed integers. If a real irrational x exists such that

$$x = \lim_{n \to \infty} p_{1n}/p_{2n} \tag{2}$$

we call x computable by additions alone. It turns out that x is computable by additions alone if and only if it is algebraic.

To prove it let us suppose first that x is so computable. Then the system (1), treated as a linear homogeneous system of first-order difference equations with constant coefficients, can be solved for the quantities p_{in} as functions of n : $p_{in} = f_i(n)$. The functions $f_i(n)$ are exponential polynomials:

$$f_i(n) = \sum_{j=1}^{s} p_{ij}(n)\lambda_j^n$$

where λ_1, λ_2, ..., λ_s are all the distinct eigenvalues of the matrix (c_{ij}) and the coefficients of the polynomials $p_{ij}(n)$ are algebraic numbers. It follows now from the hypothesis that x is the ratio of certain two such coefficients, hence it is algebraic.

Conversely, suppose that x is a real irrational algebraic number, so that it satisfies an irreducible polynomial equation $P(x) = 0$ with integer coefficients. Replacing, if necessary, $P(x)$ by $P(x)$ divided by the greatest common divisor of $P(x)$ and $P'(x)$, we may assume that x is a simple root and so are all its conjugates. Next, we show that x may be assumed to exceed all its conjugates in absolute value. For if it doesn't we can find integers a and b, with $b \neq 0$, such that

$$|x - a/b| < |u - a/b|$$

for any conjugate u of x (other than x itself). Now

$$X = \frac{1}{x - a/b}$$

also satisfies a polynomial equation with integer coefficients and it certainly exceeds all its conjugates in absolute value. Suppose that k sequences p_{1n}, $p_{2n}, \ldots, p_{kn}$, $n = 0, 1, 2, \ldots$ can be found so that X satisfies the condition (2) defining the computability by additions alone:

$$X = \lim_{n \to \infty} p_{n1}/p_{n2},$$

then

$$x = \lim_{n \to \infty} \frac{bp_{2n} + ap_{1n}}{bp_{1n}}$$

so that, adjoining two more sequences, bp_{1n} and $ap_{1n} + bp_{2n}$, to the other k sequences and renumbering the sequences we obtain $k + 2$ sequences for which (2) holds.

It remains to be shown that when x exceeds all its conjugates in absolute value then k sequences satisfying (1) and (2) can be found. Since for any fixed positive integer N the relations

$$x = \lim_{n \to \infty} p_{1n}/p_{2n}, \qquad x/N = \lim_{n \to \infty} p_{1n}/Np_{2n}$$

are equivalent, we may suppose that x is an algebraic integer. We observe that the matrix

$$\begin{pmatrix} & & 1 & & \\ & & & \cdot & \\ & & & & \cdot \\ & & & & & \cdot \\ & & & & & 1 \\ -a_k & -a_{k-1} & \cdots & & -a_1 \end{pmatrix}$$

in which all elements not shown are equal to 0, has the characteristic equation

$$\lambda^k + a_1 \lambda^{k-1} + \cdots + a_{k-1}\lambda + a_k = 0;$$

therefore the matrix $C = (c_{ij})$ can be found so that x is one of its eigenvalues and all other eigenvalues are less in absolute value. With suitable initial values $p_{10}, p_{20}, \ldots, p_{k0}$, we shall then have

$$\lim_{n \to \infty} p_{1\,n+1}/p_{1n} = x$$

so that x is computable by additions alone.

This shows that, in our sense of computing, additions alone are not enough to compute transcendental numbers. We modify our computational scheme now: the numbers $p_{1\,n+1}, \ldots, p_{k\,n+1}$ preceding equation (1) are required to be generable from their predecessors $p_{1n}, \ldots, p_{kn}$ by fixed additions and multiplications; everything else is as before. Instead of (1) we have

$$p_{i\,n+1} = P_i(p_{1n}, \ldots, p_{kn}), \qquad i = 1, 2, \ldots, k; \quad n = 0, 1, \ldots$$

where the P_i are k polynomials, each in k variables and with integer coefficients. If (2) holds we call the real irrational number x AM-computable (computable by additions and multiplications). Since x is irrational $k \geq 2$ for we need one sequence for numerators and one for denominators. It turns out that a real irrational x is algebraic if and only if $k = 2$, i.e., if the *minimum* number of sequences necessary is 2. The sufficiency of two sequences is easily shown by a suitable encoding of Newton's algorithm. To complete the proof let

$$x = \lim_{n \to \infty} p_n/q_n \tag{3}$$

where

$$p_{n+1} = P(p_n, q_n), \qquad q_{n+1} = Q(p_n, q_n), \qquad n = 0, 1, \ldots.$$

p_0 and q_0 are given and the polynomials P and Q have integer coefficients. By decomposing P and Q into sums of homogeneous polynomials P_j and Q_j of various degrees we have

$$p_{n+1}/q_{n+1} = q_n^{N-M} \sum_{j=0}^{N} q_n^{-(N-j)} P_j(p_n/q_n, 1) \Big/ \sum_{j=0}^{M} q_n^{-(M-j)} Q_j(p_n/q_n, 1);$$

hence by (3) $N = M$ and either $P_N(x, 1) = 0$ or $Q_N(x, 1) = 0$ or

$$x = P_N(x, 1)/Q_N(x, 1),$$

and so x is algebraic. The reader may wish to verify that the transcendental numbers

$$\pi, \quad e, \quad \log 2, \quad \sum_{n=0}^{\infty} 2^{-2^n}$$

are all of them three-sequence numbers. No example of a number which requires minimally more than three sequences appears to be known. It may be conjectured that if $p_1, p_2, \ldots$ are the primes 2, 3, 5, $\ldots$ then the number

$$\sum_{n=0}^{\infty} \left(\sum_{i=1}^{k} p_i^{p_i^n} \right)^{-1}$$

requires exactly $k + 2$ sequences.

3. POLYGONAL PATHS AND MINIMIZATION

It is easy to prove that n points in the plane, no three of them collinear, are the vertices of a non-self-intersecting n-gon. We prove this in a way which introduces some structure and appears to be adapted to generalizations. Divide the n points $P = \{p_1, p_2, \ldots, p_n\}$ into layers as follows: the first layer P_1 consists of the vertices of the convex hull of P, the second layer of the vertices of the convex hull of $P - P_1$, and so on. We can visualize the layers by driving vertical pins into the plane at $p_1, p_2, \ldots, p_n$ and putting a perfectly elastic rubber band round the pins. The band rests on certain pins which form the first layer P_1; when we remove them it will rest on the pins of the second layer, and so on.

We obtain thus k layers $P_1, \ldots, P_k$, with $1 \leq k \leq 1 + n/3$, and the points of each layer are vertices of a convex polygon. By removing and adding certain edges it is now easy to splice the k convex polygons so as to prove our proposition. Generalizations to three or more dimensions may be handled similarly though the details are harder. The reader may wish to consider also the generalization to infinitely many points.

In the plane case there will be in general many non-self-intersecting polygons which have the assigned n vertices, and we may ask which of these polygons minimizes a certain given quantity, for instance, which one surrounds the smallest area or has the minimum circumference. The latter is the concrete case of the so-called problem of the traveling salesman: If the salesman has n places (supposed to be points) to visit, how is a round trip to be arranged to minimize the total distance traveled? To prove that this is indeed so, that our problem *is* the concrete case of the traveling salesman problem, we must show that the minimal-length closed path which passes through every point cannot be self-intersecting. With reference to Figure 1a we note that

$$ac + db = dx + xc + ax + xb;$$

it follows that the sum of the lengths of the diagonals exceeds the sum of the lengths of either pair of nonadjoining sides. Hence the minimal length path cannot be self-intersecting as shown in Figure 1b for it gets shortened by removing db and ac, and adding ad and bc. We note the general form of the traveling salesman problem. For that purpose we consider n points $p_1, \ldots, p_n$ and all the $2\binom{n}{2}$ ordered pairs (p_i, p_j), $i \neq j$, called directed edges; to the edge (p_i, p_j) we assign the cost c_{ij} which is a real number not necessarily positive; also c_{ij} is not necessarily equal to c_{ji}. A Hamiltonian circuit is a closed path consisting of coherently directed edges, which visits every point p_i just once and its cost is the sum of the costs of its edges. Now the problem is to find the cheapest Hamiltonian circuit.

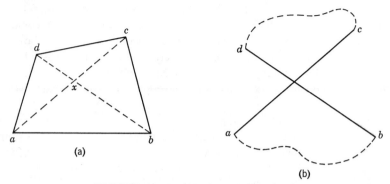

FIGURE 1.   Lengths in a quadrilateral.

For comparison purposes we introduce three further problems concerned with path minimization for n points $p_1, \ldots, p_n$ in the plane. In the minimum distance problem we ask for the point x which minimizes the length sum $\sum_1^n p_i x$. In the problem of the shortest spanning subtree we ask for a subset of the $\binom{n}{2}$ undirected straight edges $p_i p_j$, in which every pair of points p_i, p_j is connected by a sequence of edges, and which is of shortest overall length. In the Steiner problem we are free to introduce any number of additional points $s_1, \ldots, s_k$ and we then ask for the shortest spanning subtree of the set of $n + k$ points $p_1, \ldots, p_n, s_1, \ldots, s_k$.

It will be observed that both forms of the traveling salesman problem as well as the shortest spanning subtree problem are discrete. That is, in each problem the number of possible candidates for solution is finite (though possibly very large for large n). Hence in each case an *effective* algorithm can be given simply by examining all possibilities. What we really want here is an *efficient* or *practicable* algorithm, i.e., one which uses sufficiently few elementary operations for the problem to be solvable for reasonably large n in a reasonably short time, say, on a large-scale computer. No such algorithm exists as yet for the traveling salesman problems; the best solutions here take about 2^n operations. It is even conjectured that no better algorithm is possible.

On the other hand, a perfectly practicable algorithm for the shortest spanning subtree has been devised [59]: we start with any point p_i, and we join it to the p_j which is closest to p_i, the segment $p_i p_j$ is called the first fragment and it is joined to the p_k which is closest to the first fragment by the corresponding segment yielding the second fragment, which is in turn joined to the nearest point to it, etc. The total number of operations is of the order of n^2. The same technique works for the generalization of the shortest spanning subtree problem, analogous to the generalization of the traveling salesman problem and depending on the introduction of a cost c_{ij} corresponding to the (undirected) connection of p_i to p_j.

The Steiner problem appears to have been first posed, and solved, for $n = 3$ by B. Cavalieri (1598–1647), an Italian geometer who was a precursor of Newton and Leibniz on the subject of integral calculus. Sometimes, however, the statement of the problem (for $n = 3$) is credited (as in [20]) to P. Fermat (1601–1665) and its solution to E. Torricelli (1608–1647), an assistant to Galileo, known for his work on motion, hydrodynamics, and mercury barometers. The case of general n appears to be due to Steiner.

On account of the k additional points $s_1, \ldots, s_k$ the problem depends, prima facie, on $2k$ continuous parameters (the coordinates of the points s_j). A simple combinatorial argument, based on counting one thing in two different ways, shows that $k \leq n - 2$. First, since the graph corresponding to the problem is a tree with $n + k$ vertices, the total number of edges is $n + k - 1$, as shown in Section 9. Suppose that among the n vertices there are v_1 with one edge, v_2 with two, and v_3 with three edges, so that $v_1 + v_2 + v_3 = n$. Counting the edges by the vertices from which they issue, we find that the number of edges is

$$(3k + v_1 + 2v_2 + 3v_3)/2;$$

the denominator 2 occurs because each edge is now counted twice. Equating the above to $n + k - 1$, we find

$$n - k - 2 = v_2 + 2v_3$$

so that $k \leq n - 2$, since $v_2 \geq 0$, $v_3 \geq 0$. We shall now reduce the Steiner problem to a large-size discrete problem with a finite number of candidates for solution, thereby providing an effective though highly inefficient algorithm for solving it. Moreover, it turns out that the solution can be found by ruler-and-compasses geometrical constructions.

We consider first some special cases. For $n = 3$ we have $k = 0$ or $k = 1$ depending on whether the triangle $p_1 p_2 p_3$ does or does not have an angle $\geq 120°$. The solutions for the two cases are illustrated in Figure 2. For the case of Figure 2b the three equilateral triangles are built outward on the sides of the triangle $p_1 p_2 p_3$, and the reader may wish to show that: (a) the three segments $s_1 p_1$, $s_1 p_2$, $s_1 p_3$ subtend at s_1 angles of 120°, (b) any four points p_i, p_j, p_{ij}, s_1 lie on a circle, and (c) the length $L = s_1 p_1 + s_1 p_2 + s_1 p_3$ of the shortest connecting network satisfies

$$L = p_1 p_{23} = p_2 p_{13} = p_3 p_{12}.$$

A part of this theorem is credited to Napoleon Bonaparte, the emperor of the French. Besides being a geometer in a small way, Napoleon was a great patron of mathematics and mathematicians. He is supposed to have said that the condition of a nation is closely related to the condition of its mathematics; he was a supporter of Gaspard Monge (1746–1818), one of the founders of

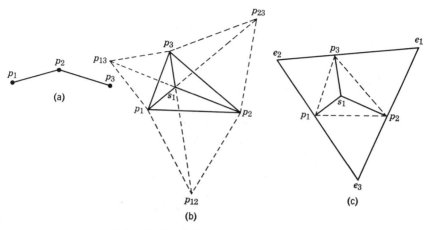

FIGURE 2.   Steiner's problem for $n = 3$.

both Ecole Normale and Ecole Polytechnique (these institutions have pioneered modern curriculum in sciences, mathematics, and engineering). According to a story, Napoleon gave the ministry of the interior to P. S. Laplace, and took it away from Laplace half a year later on account of Laplace's incompetence, with the words "he brings into the affairs the spirit of the infinitely small."

With reference to Figure 2c the reader may wish to supply an alternative proof of the minimal-length property of the Steiner point s_1, based on the theorem of V. Viviani (1622–1703). Let $e_1 e_2 e_3$ be the equilateral triangle whose sides pass through p_1, p_2, p_3 as shown; the minimal-length property of s_1 follows from Viviani's theorem which asserts that for an arbitrary point x

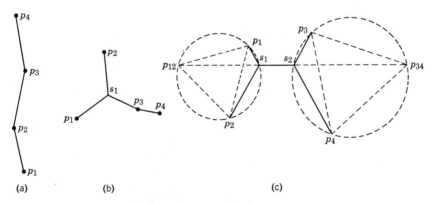

FIGURE 3.   Steiner's problem for $n = 4$.

inside $p_1 p_2 p_3$ the sum of its distances to the sides of $e_1 e_2 e_3$ is a constant independent of x. For $n = 4$ we have $k = 0$, $k = 1$, or $k = 2$; the cases $k = 0$ and $k = 1$ are illustrated in Figure 3a and b. In the case $k = 2$ illustrated in Figure 3c we first find by the equilateral construction the points p_{12} and p_{34}, and we join them by a straight segment. Then we locate s_1 and s_2 on $p_{12} p_{34}$ at the intersections with the shown circles. There is actually another candidate for the shortest connecting network obtained as in Figure 2c but by pairing p_1 with p_3 and p_2 with p_4. In the case illustrated the total length of the connecting network is $p_{12} p_{34}$, and since $p_{12} p_{34} < p_{13} p_{24}$ the solution is as shown. For the special case $n = 10$ illustrated in Figure 4 we have two components: p_1, p_2, p_3, p_4, p_5 and p_6, p_7, p_8, and two residual points p_9 and p_{10}. For each component considered as a separate initial set, there is the maximum possible number of s-points: $5 - 2 = 3$ for the first one and $3 - 2 = 1$ for the second one.

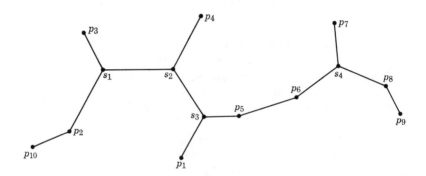

FIGURE 4.   Steiner's problem for $n = 10$.

We consider now the general case of n points. Here, too, we have a subdivision into some q components, each of at least three points, and a residual set of r points. We observe that a point p_i can belong to at most three components and that any two components have at most one p_i-point in common. Thus, though we do not know a priori how to divide the given n points into components and the residual, the total number of possible subdivisions is finite. Therefore it suffices to show how to solve Steiner's problem for a single component, and the rest follows, as in the shortest spanning subtree problem.

Over one component we solve Steiner's problem inductively. Let $p_1, \ldots, p_m$ be the m points of the component; we certainly know how to solve the problem for $m = 3$. Suppose that we also know how to solve it for $3, 4, \ldots, m - 1$ points. With reference to Figure 5 we erase the points p_1 and p_2 and we add the point p_{12}, noting that p_{12}, s_1, s_2 are collinear. We do not know whether p_1 is correctly paired with p_2; it might have to be paired instead with another p_i. Thus there are $2(m-1)$ candidates for the hypothetical p_{12}. This gives us $2(m-1)$ systems of $m-1$ points to solve the Steiner problem for, and therefore the induction is complete.

We turn now to the minimum distance problem; for $n = 3$ this coincides with Steiner's problem. Given the n points $p_1, \ldots, p_n$ let x be the point minimizing the sum $\sum_1^n p_i x$ and let L be any straight line in the plane, disjoint from all the p_i's and separating some k of them, say $p_1, \ldots, p_k$, from the $n-k$ others, $p_{k+1}, \ldots, p_n$. We now define polyellipses: let E_1 be the k-ellipse with foci $p_1, \ldots, p_k$ passing through x, i.e., the locus of points in the plane the sum of whose distances from $p_1, \ldots, p_k$ is constant; similarly let E_2 be the $(n-k)$-ellipse through x with foci $p_{k+1}, \ldots, p_n$. Then, by the minimum property of x, E_1 and E_2 are externally tangent at x. In particular, for $n = 3$ and $k = 1$ we

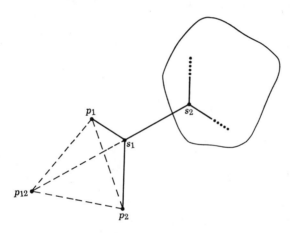

FIGURE 5. Single component in Steiner's problem.

have an ordinary ellipse tangent to a circle, and we have a simple proof of the $120°$ angle property for Steiner's problem. In the general case the tangency property enables us to develop some fast-convergent algorithms for the determination of x. Also, generalizing k-ellipses to k-ellipsoids, we can prove some results for higher dimensions, for instance, the equivalent of the $120°$ angle property for tetrahedra and simplexes.

Since for $n = 3$ the minimum distance and Steiner problems coincide, x can be determined for $n = 3$ by ruler and compasses. For $n = 4$ there are two cases to consider: either the four points p_i are the vertices of a convex quadrilateral, or one of them lies inside the triangle determined by the other three.

In the first case let x be the intersection of the diagonals of the quadrilateral as shown in Figure 6, and let u be some other point. Then

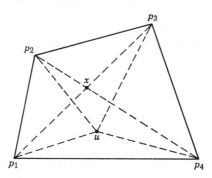

FIGURE 6. Minimum distance problem for $n = 4$.

$$p_1 u + p_3 u > p_1 p_3 = p_1 x + p_3 x, \qquad p_2 u + p_4 u > p_2 p_4 = p_2 x + p_4 x$$

and hence

$$\sum_1^4 p_i u > \sum_1^4 p_i x;$$

this proves the minimum property of x. In the second case, when some p_i is inside the triangle determined by the three others, the previously mentioned tangency property may be used to show that the minimum of $\sum_1^4 p_i x$ is attained when x is the internal p_i.

On the other hand, for $n \geq 5$ the minimum distance problem is in general not solvable by ruler-and-compasses geometrical constructions. The proof provides an exercise in elementary Galois theory. We take the five points p_i to be at $(0, 1)$, $(0, 0)$, $(0, -1)$, $(3, 3)$, $(3, -3)$; by symmetry the minimum distance point is at $(y, 0)$. Finding the derivative and equating it to 0 we find that y satisfies an octic equation with integer coefficients; reducing modulo 11 we find that the octic polynomial has two irreducible factors: a cubic and a quintic. It follows that the order of the Galois group of the original octic over the rationals is divisible by 15; thus it cannot be a power of 2 and so the octic is not solvable by radicals; this proves the assertion.

We observe that both the minimum distance problem and the Steiner problem for a single component can be solved by simple mechanical analogues shown in Figure 7a and b. In each case threads are tied together to small rings as shown, led over small rotatable pulleys and then through the holes in the boards (pegboards could be used), and equal weights are tied to the threads,

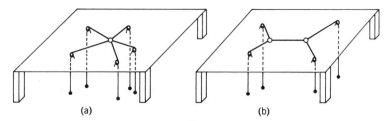

FIGURE 7. Mechanical analogues.

The minimum potential energy requirement shows that the linear configurations on top of the boards have minimal length. The reader may wish to consider the effect of using unequal weights.

4. SOME TYPES OF SYMMETRY

An electric circuit consists of a battery and twelve 1-ohm resistances forming the edges of a cube as shown in Figure 1a. What is the resistance between the points a and h? By rotating the cube about the line ah through 120° and 240° we find by symmetry that the points c, b, d are at the same potential, and so are the points e, f, g. Therefore the circuit of Figure 1a is equivalent to that of Figure 1b and the resistance between a and h is 5/6 ohm. The reader may wish to solve the analogous problems for the regular octahedron and icosahedron as well as for various higher-dimensional analogues. It is also possible to use symmetry in calculating the resistance between other pairs of points of a cubical or other network. It may be of interest to find out when the equivalent circuits are series-parallel; in the case of a cube there are three distinct pairs of terminals: a–h, a–f, a–b and it turns out that the equivalent circuits of a–h and a–b are series-parallel while that of a–f is not.

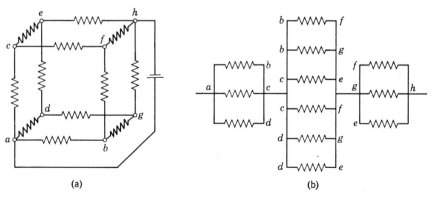

FIGURE 1. Cubical network.

Similar symmetry argument will help in solving a seemingly unrelated problem in solid geometry: two opposite vertices a and h of a cube are joined by a straight segment ah, the plane P is perpendicular to ah and bisects it; in what polygon Q does P cut the cube? By a count of edges it is easy to show that Q is a hexagon; moreover, all its sides are equal. Let the angles of Q be, in order, $\alpha_1, \alpha_2, \alpha_3, \alpha_4, \alpha_5, \alpha_6$. Rotating the cube by $120°$ and $240°$ about ah we find that $\alpha_1 = \alpha_3 = \alpha_5$, $\alpha_2 = \alpha_4 = \alpha_6$. To show that all angles are equal we need another symmetry: reflecting the cube in the plane through a, c, h, g. It follows now that all angles are equal, and so Q is a regular hexagon. The reader may wish to generalize this to n dimensions, in particular, to the four dimensional cube. In the general case of an n-cube the reader may wish to find the various types of sections of the n-cube by a hyperplane orthogonal to the long diagonal ah, and moving from a to h. The preceding examples illustrate discrete rotational symmetry. We have in each case a geometrical configuration C with a distinguished straight line L; if R denotes a rotation of C about L through a suitable angle $\alpha = 2\pi/n$ then in the language of group theory the rotations $I, R, R^2, \ldots, R^{n-1}$ form a group of symmetries of C. I is the identity, and for both problems about the three-dimensional cube $n = 3$. A simple use of discrete translational symmetry is given in the chapter on series and products where the function

$$\sec^2 x = \sum_{n=-\infty}^{\infty} (x - n\pi)^{-2}$$

is shown to have period π because on replacing x by $x + \pi$ each term in the series is changed into the next one. In the language of group theory we can put it thus: on the real line consider the infinite discrete set of points

$$S = \{.., (-2\pi), (-\pi), (0), (\pi), (2\pi), \ldots\}$$

and let T denote the translation by π to the right; then S is invariant under $T^{\pm n}$ for any integer n. Therefore, here the group of symmetries is infinite cyclic with one generator T.

We illustrate next—on the examples of two simple problems—two other types of symmetry, spiral and helical. Both cases will deal with continuous rather than discrete symmetry. For our first example let n points $p_1, \ldots, p_n$ ($n \geq 3$) be located at the time $t = 0$ at the consecutive vertices of a regular n-gon P of side-length a. The points then begin to move with the same constant speed v and in such a manner that p_i aims at p_{i+1} throughout the motion (here $i = 1, \ldots, n$ and $p_{n+1} \equiv p_1$). The motion continues till all points meet at the center c of P; without using calculus we are to find the distance d traversed by each point. Let $p_i(t)$ be the position of p_i at the time t. Then the symmetry of the configuration shows that $p_1(t), \ldots, p_n(t)$ form the n vertices of a regular n-gon $P(t)$ which shrinks and rotates about c with increasing t, finally collapsing to the single point c. It follows that the angle between the straight segment from $p_i(t)$ to c, and the instantaneous direction of motion of $p_i(t)$ (i.e., toward $p_{i+1}(t)$) is constant $\alpha = (n - 2)\pi/2n$. Hence it follows that $p_i(t)$ approaches c with constant *radial* speed $v \cos \alpha$ and it is now not hard to show that $d = a/2 \sec^2(n - 2)\pi/2n$. It may be added that the speed v need not be constant (so long as it is the same for all p_i's at all times t).

The reader may observe that if we rewrite $p_i(t)$ as $z_i(t)$, treating it as a complex number depending on the real parameter t, with c as the origin of the complex plane, then a complex number b exists such that

$$P(t) = e^{bt}P(0).$$

Thus the group of symmetries here is the multiplicative group of complex numbers $\{e^{bt}\}$, and b plays the role of the infinitesimal generator (see Section 9 of the chapter on series and products). The trajectory of a point under this group is the logarithmic spiral with polar equation $r = Ae^{k\theta}$, hence the name spiral symmetry.

For our second example we take a circular helix H in the three-dimensional Euclidean space, with the parametric equations

$$x(t) = a \cos t, \qquad y(t) = a \sin t, \qquad z(t) = bt.$$

We wish to show that the radii of curvature and of torsion of H are constant, and that the locus of centers of curvature of H is a helix L coaxial to H. We shall find L by using the helical symmetry of H: if H undergoes a suitable screw-motion then it just slides along itself; the rest of the problem will fall out automatically. As a matter of fact we do not even have to know what the center of curvature is but merely that it is *locally definable*. Here, for instance, together with each point $p = p(t)$ of H we consider two "near-by" points of H: $q = p(t + \Delta t)$ and $r = p(t - \Delta t)$; let $c(t, \Delta t)$ be the center of the circle circumscribing the triangle pqr; then the center of curvature of H at $p(t)$ is the point

$$c(t) = \lim_{\Delta t \to 0} c(t, \Delta t).$$

If now $p(t_1)$ is another point of H then the screw-motion of H along itself, which carries $p(t)$ onto $p(t_1)$, will also carry $c(t)$ onto $c(t_1)$. Hence the locus L of all points $c(t)$ is a helix coaxial to H.

We finish this section with an example due to Sierpinski [69]. In all previous cases we were concerned with symmetries of geometrical configurations, that is, with motions (or more general transformations as in the spiral case) which carry the configurations into themselves. In Sierpinski's example we are interested in *lack of symmetry* and our aim is reverse: to construct a configuration which does not coincide with itself under certain transformations. This lack of symmetry is reflected in a property of certain algebraic structure of transformations: it is *free* (more precisely, a free semigroup). The problem here is: to construct a plane set of points X which can be decomposed into two disjoint parts A and B each of which is congruent to the whole of X be a rigid motion. In the Cartesian plane we let o be the origin, Tu the result of translating a point u by one unit in the positive x-direction, and Ru the result of rotating u by one radian counterclockwise about o. Consider now the set X of all points which can be obtained by applying T and R to o any finite number of times and in any order; this construction may be systematized in the tree-diagram of Figure 2. More formally, we say that X is obtained by acting on

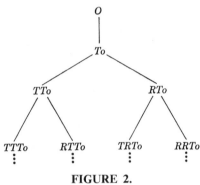

FIGURE 2.

o with the semigroup H generated by R and T. Since π is transcendental and so are cos 1 and sin 1, it may be shown that no two points of X coincide (which means that H is free); the reader may wish to show that. Next, X is a disjoint union of A and B, where A consists of all strings such as $TR \dots To$ which start with T, and B consists of o and all strings starting with R. Now we have $A = TX$ and $B = RX$ and the problem is finished.

It may be remarked that a somewhat similar construction is at the basis of the so-called paradoxical decomposition of the two-dimensional sphere S: S is a disjoint union of two parts A and B each of which is equivalent to S by finite decomposition (see also Section 7 of Chapter 1). Unlike X which is unbounded, S is bounded; it may be shown that such a paradoxical decomposition cannot occur in the plane for a bounded set; the reason behind this is of algebraic nature: certain free groups do not exist for plane motions.

5. EXTREMAL PROBLEMS AND EXTREMAL POINTS

What is the largest volume of a tetrahedron T which is contained in a right circular cylinder C of height h and base-radius a? This elementary problem illustrates a number of simple but useful points in analysis and geometry. One of these is the principle of minimum variation: in maximizing or minimizing a quantity f which depends on several variables we investigate the effect of perturbations of f in which possibly few variables change. For instance, let us prove that of all n-gons inscribed in a circle the regular one has the largest area. Consider any three consecutive vertices x_i, x_{i+1}, x_{i+2} and let x_{i+1} vary on the circle while all $n-1$ others are fixed. The only variation in the area of the polygon arises from the triangle $x_i x_{i+1} x_{i+2}$ whose area is largest when x_{i+1} is furthest from the line $x_i x_{i+2}$. Hence $x_i x_{i+1} = x_{i+1} x_{i+2}$ and by a symmetric argument all sides of the polygon are equal.

We return to our main problem. Let the vertices of T be v_1, v_2, v_3, v_4; then all four lie in the boundary of C, for otherwise a vertex v could be moved inside C away from the plane of the other three vertices, thereby increasing the volume of T. Next, suppose that a vertex, say v_1, lies in the lateral part of the boundary of C but not on either of its circular rims, as shown in Figure 1a. Let P be the plane through v_2, v_3, v_4. P is either parallel to the axis of C or not. In the first case v_1 may be moved up or down without changing the volume of T. In the second case moving v_1 in one direction (up or down) increases the volume of T and moving it in the opposite direction (down or up) decreases it. Similar argument applies when a vertex lies in the base or the top of C but, again, not on the rim of either.

Hence we may assume that each vertex of T is an extremal point of the boundary of C: this means that if a straight segment in the boundary passes through the point in question then that point must be an end-point of such a segment. However, the extremal points in our case make up just the upper and lower rims. Hence the vertices lie in the two rims and we have two possibilities: three vertices on one rim and the fourth on the other (Figure 1b), or two vertices on each rim (Figure 1c).

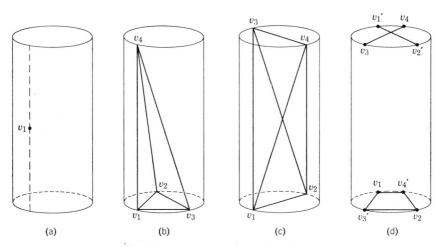

(a)   (b)   (c)   (d)

FIGURE 1. Tetrahedron and cylinder.

In the first case the volume V of T depends only on the area of the triangle $v_1 v_2 v_3$ which, as we have shown, must be equilateral for the maximum; hence in this case the (hypothetical) maximum is $V = 3^{1/2}a^2h/4$. To be able to handle the second case in the same way we invoke the Cavalieri principle: let S_1 and S_2 be two solids; if there exists a direction u such that every plane perpendicular to u cuts S_1 and S_2 in plane figures of the same area, then S_1 and S_2 have the same volumes. With reference to Figures 1c and 1d, we project v_1 and v_2 onto the upper rim as v_1' and v_2', and v_3 and v_4 onto the lower rim as v_3' and v_4'. Let Q be the quadrilateral $v_1 v_4' v_2 v_3'$, and let v be the intersection of the straight segments $v_3 v_4$ and $v_1' v_2'$. Suitable use of Cavalieri's principle will show that T has the same volume as the pyramid B based on Q with the vertex v. The volume of B depends again only on the area of Q: hence for a maximum Q is a square, and so the volumes of B and T are here $2a^2h/3$. Since $2/3 > 3^{1/2}/4$ we find that the maximum volume of T is $2a^2h/3$ and the corresponding maximal tetrahedron has its edges $v_1 v_2$ and $v_3 v_4$ perpendicular and each a diameter of the circular rim.

The reader may wish to find out whether the other candidate is a local maximum, and to see whether two polyhedra or other solids may have several Cavalieri directions.

Our next problem is really a variational one, since here the extremum depends on the whole course of a curve; the reader may wish to compare this with Section 7 of Chapter 1. A man in a boat, at a unit distance from a straight shore, finds himself lost in a completely impenetrable fog without knowing the direction of the shore. What is the shortest sailing curve the boat should follow, to make sure of hitting the shore? Mathematically, the problem is this: what is the shortest arc C in the plane, starting from the origin o and having the property ($C1$): C has a point in common with every straight line in the plane at a unit distance from o? There is, of course, no a priori guarantee that such a minimal arc exists; we should really speak of infima instead of minima.

We observe first that an obviously admissible sailing curve C_1 is the one shown in Figure 2a; here the boat moves one unit in some direction and then describes the unit circle about its original position o. It is possible to show, and the reader may wish to show it, that C_1 is the minimal-length solution to another problem, forming a simple variant of ours, in which the shore line is arbitrary rather than straight. We shall now use what may be called the principle of small perturbations of different orders of magnitude, to show that C_1 is not minimal for our problem. For that purpose we consider the curve C_2 of Figure 2b, consisting of a straight segment and an arc of a circle about o. Its length is

$$L(\varepsilon) = (2\pi + 1 - 2\varepsilon)\sec \varepsilon \tag{1}$$

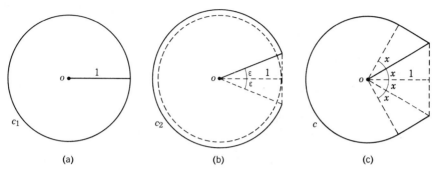

(a) (b) (c)

FIGURE 2. Minimal sailing curve.

while the length of C_1 is $L(0) = 2\pi + 1$. Hence

$$\frac{L(\varepsilon)}{L(0)} = \left(1 - \frac{2}{2\pi + 1}\,\varepsilon\right)\sec\varepsilon$$

$$= \left(1 - \frac{2}{2\pi + 1}\,\varepsilon\right)\left(1 + \frac{\varepsilon^2}{2} + 0(\varepsilon^3)\right) = 1 - \frac{2}{2\pi + 1}\,\varepsilon + 0(\varepsilon^2);$$

thus it follows that for sufficiently small $\varepsilon > 0$ we have $L(\varepsilon) < L(0)$ and so C_1 is not minimal.

Similar argument may be used in other extremum problems. An example is the volume-edge length isoperimetry for convex polyhedra: to find a convex polyhedron H, of maximum volume among all convex polyhedra whose sum of edgelengths is kept constant. Let $V(H)$ and $L(H)$ be the volume and the sum of lengths of all edges, of a convex polyhedron H; we wish to determine H so that

$$\frac{V(H)}{L^3(H)} = \max.$$

It is apparently as yet unknown what the maximizing H is, though it may be conjectured that it is the right prism based on an equilateral triangle and bounded laterally by three squares. However, we can show that the maximal polyhedron H has the property that not too many walls can meet at any of its vertices. We observe first a simple lemma in plane geometry which the reader may wish to prove (and perhaps generalize): there is an integer k such that if A is any convex n-gon and $n \geq k$, then for a suitable point inside A the sum of distances to the vertices is not less than the circumference of A. We can also express this as follows. Let A be an arbitrary convex n-gon with the vertices enumerated consecutively $v_1, v_2, \ldots, v_n$; then there is an integer k such that if $n \geq k$ then

$$\max_{x \in A} \sum_{i=1}^{n} |x v_i| \geq \sum_{i=1}^{n} |v_i v_{i+1}|, \qquad v_{n+1} \equiv v_1.$$

Here $x \in A$ means that the point x is inside A.

Suppose now that the maximal polyhedron H has a vertex v at which k or more walls meet. Let w be a point inside H, such that $|vw| = \varepsilon$ is sufficiently small, and let P be the plane through w, orthogonal to vw. Let H_ε be the polyhedron obtained by slicing off H the small piece based on P. By the geometrical lemma quoted above we can arrange the position of w so that

$$L(H) > L(H_\varepsilon), \qquad L(H) - L(H_\varepsilon) = K_1 \varepsilon, \qquad V(H) - V(H_\varepsilon) = K_2 \varepsilon^3$$

with positive K_1 and K_2. Hence

$$\frac{V(H_\varepsilon)}{L^3(H_\varepsilon)} = \frac{V(H)}{L^3(H)} \left[1 + \frac{3K_1}{L(H)} \varepsilon + 0(\varepsilon^2) \right]$$

and so $V(H_\varepsilon)/L^3(H_\varepsilon)$ exceeds $V(H)/L^3(H)$ for sufficiently small ε; therefore a convex polyhedron H with k or more walls meeting at a vertex cannot be maximal.

We return now to our problem of the shortest sailing curve. Consider the curve shown in Figure 3a, from o to c; this could not possibly be the shortest sailing curve for if we replace the arcs $\widehat{pq}$ and $\widehat{rs}$ by the corresponding straight segments pq and rs the new curve still has the property (C1) if the original one had it, and the new one is shorter. This shows that the shortest sailing curve has spiral shape: unit tangent vector turns, though possibly discontinuously, always in the same direction (alternatively: any small subarc must be convex). For the same reason, the minimal curve cannot have subarcs $\widehat{pq}$ or $\widehat{rs}$ whose extended secants pq or rs have no points in common with the

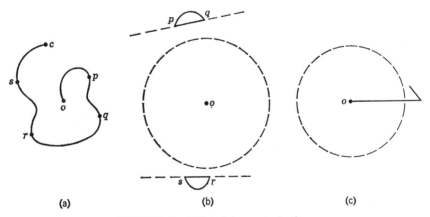

(a)   (b)   (c)

FIGURE 3.   Minimizing perturbations.

unit circle about o, as shown in Figure 3b, For, we simply replace $\widehat{pq}$ and $\widehat{rs}$ by pq and rs. Similar argument shows that the initial part of the minimal curve, inside the unit circle about o, is a straight segment. Further, another application of the principle of small perturbations of different order shows that this initial straight segment continues beyond the unit circle, as shown in Figure 3c.

It is now not hard to show that the minimal curve C must be as shown in Figure 2c. Its length is

$$L(x) = 2\pi - 4x + 2 \tan x + \sec x$$

and so for a minimum we have

$$\sec x = \left(\frac{15 - 33^{1/2}}{6}\right)^{1/2} \simeq 1.242, \qquad x \simeq 0.6349;$$

hence the length of the shortest sailing curve is 6.459 (while the length of the admissible curve C_1 of Figure 2a is 7.283).

The reader may wish to consider the analogous minimal sailing curve problem for a boat lost somewhere in a straight canal of unit width. Also, our problem may be generalized to three or more dimensions. The general n-dimensional problem is: in the n-dimensional Euclidean space E^n find the shortest arc starting from the origin o, which has a point in common with every plane at a unit distance from o. For $n = 3$ the reader may consider the following initial candidate C_1 for the minimal arc C; let v_1, v_2, v_3, v_4 be the vertices of a regular tetrahedron whose center is o, and whose inscribed sphere has unit radius; now C_1 is as shown in Figure 4. However, it is possible that one may save on length by "cutting corners" in C_1. We finish by computing an upper bound for the length L_n of the minimal sailing curve in E^n. It is easy to show that the n-dimensional equivalent $ov_1v_2 \ldots v_{n+1}$ of the curve of Figure 4 is an admissible sailing curve in E^n. Let $u_1, u_2, \ldots, u_{n+1}$ be the n unit

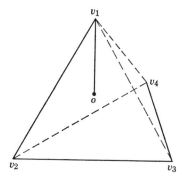

FIGURE 4. An economical sailing curve in E^3.

vectors originating at o and pointing toward the $n + 1$ vertices of the regular simplex. By symmetry we have

$$u_i \cdot u_i = 1, \qquad u_i \cdot u_j = c \qquad \text{for all } i, j, \text{ if } i \neq j,$$

$$\sum_{i=1}^{n+1} u_i = 0.$$

Hence, by multiplying out in

$$\sum_{i=1}^{n+1} u_i \cdot \sum_{i=1}^{n+1} u_i = 0$$

we have $c = -1/n$. Next, $ov_i = nu_i$ since our regular simplex has the unit sphere as its inscribed sphere, and since in any n-dimensional simplex (regular or not) the $n + 1$ medians pass through one point which divides each median in the ratio $n : 1$. It is now not hard to compute the length of the path $ov_1v_2 \ldots v_{n+1}$ and to show hence that

$$L_n \leq n[1 + \sqrt{2n(n + 1)}].$$

6. CIRCLES ON A TORUS

How many circles lying entirely on a torus T can be drawn through an arbitrary point p of T? Assuming the torus situated horizontally we have two obvious such circles: a vertical meridional circle, and a horizontal latitude circle. However, the answer to our question is four, not two, circles. Let the torus T be generated by revolving a circle C of radius r about an axis whose distance from the center of C is a; we suppose that $a > r$. Let the coordinate systems xyz and $x_1y_1z_1$ be as shown in Figure 1; the y-axis and y_1-axis coincide and $\sin \theta = r/a$. The Cartesian equations of T are

$$\left(\sqrt{x^2 + y^2} - a\right)^2 + z^2 = r^2 \qquad\qquad\qquad xyz\text{-system,}$$

$$\left(\sqrt{(x_1 \cos \theta - z_1 \sin \theta)^2 + y_1{}^2} - a\right)^2 + (x_1 \sin \theta + z_1 \cos \theta)^2 = r^2$$
$$x_1y_1z_1\text{-system.}$$

Therefore the section of T by the x_1y_1-plane has the equation

$$\left(\sqrt{x_1{}^2 \cos^2 \theta + y_1{}^2} - a\right)^2 + x_1{}^2 \sin^2 \theta = r^2$$

which may be written as

$$[x_1{}^2 + (y_1 - r)^2 - a^2][x_1{}^2 + (y_1 + r)^2 - a^2] = 0.$$

Hence this section consists of two circles of radius a, with centers $2r$ apart. By sliding either circle round the torus once we find that it passes through every point of T just once.

We have shown that there are at least four circles in T, passing through an arbitrary point p of T; the reader may wish to complete the solution by show-

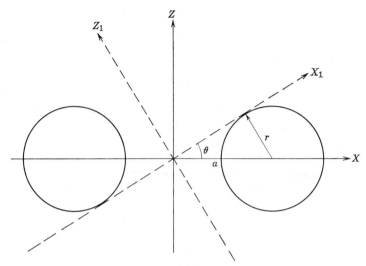

FIGURE 1.   Circles on a torus.

ing that there are exactly four. The reader may also wish to tackle the much harder problem of characterizing the torus by the four-circle property: (a) if S is a complete sufficiently smooth surface containing exactly four circles, through any point of it, then S is a torus; (b) if the number of such circles is ≥ 5, then it is infinite and the surface is a sphere.

7. COMPOSITE ESTIMATES

In some mathematical problems it is necessary to produce a good, that is, a sufficiently low, upper estimate of some quantity. A sufficiently good result may be obtained by the multiple-estimate technique. One form that this may take is in the simple and well-known problem in which a convergent sequence $a_1, a_2, \ldots$ with limit L is given and we are required to show that the sequence of the arithmetic means $m_1, m_2, \ldots$ also converges to L. We have

$$m_n = \frac{1}{n} \sum_1^n a_j, \qquad n = 1, 2, \ldots$$

and we wish to show that $m_n - L$ is small in absolute value for large n. Since

$$m_n - L = \frac{1}{n} \sum_1^n (a_j - L)$$

we prepare for the use of double estimate by observing that

$$|m_n - L| \leq \frac{1}{n} \sum_1^{N(n)} |a_j - L| + \frac{1}{n} \sum_{N(n)+1}^n |a_j - L|. \tag{1}$$

Here $N(n)$ is an integer-valued function which tends to infinity *slower* than n, its precise nature does not otherwise matter. We estimate the two terms on the right hand side of (1) differently. The first one is small for large n because: (a) $N(n)/n$ approaches 0, and (b) a_j, hence also $|a_j - L|$, is bounded. The second term is small for large n because a_j tends to limit L. To put it briefly, the terms of the first sum are possibly large but few, the terms of the second are many but small. Putting together the two estimates we show that $|m_n - L|$ is arbitrarily small for n large enough, hence $m_n \to L$. This is the basis of the Cesaro summability method; a sequence $a_1, a_2, \ldots$ is said to be C-summable to L if the sequence $m_1, m_2, \ldots$ of arithmetic means converges to L. An obvious consistency requirement for the Cesaro (or any other similar) method is that a convergent sequence ought to be summable to the value equal to its limit. This is assured by the foregoing proof.

Another example of a frequent use of a composite estimate is in connection with uniform convergence. If, for example, we wish to show that a uniformly convergent sequence $f_1(x), f_2(x), \ldots$ of continuous functions converges to a continuous limit $L(x)$, we have to estimate $|L(x + \Delta x) - L(x)|$ for small Δx. We write

$$L(x + \Delta x) - L(x) = L(x + \Delta x)$$
$$- f_n(x + \Delta x) + f_n(x + \Delta x) - f_n(x) + f_n(x) - L(x)$$

so that

$$|L(x + \Delta x) - L(x)| \leq |L(x + \Delta x) - f_n(x + \Delta x)|$$
$$+ |f_n(x + \Delta x) - f_n(x)| + |f_n(x) - L(x)|. \quad (2)$$

This tripartite appearance is characteristic of problems bearing on uniform convergence and continuity. We estimate the three terms on the right hand side of (2) as follows: the first and the last are small because $f_n(x)$ converges to $L(x)$, and therefore each is $< \varepsilon/3$ for arbitrary $\varepsilon > 0$ provided that n is large enough (since the convergence is uniform the n_0 implicit in the phrase "n is large enough" does not depend on x); the middle term can be made $< \varepsilon/3$ since each $f_n(x)$ is continuous. It follows that $|L(x + \Delta x) - L(x)|$ can be made arbitrarily small if Δx is small enough, hence $L(x)$ is continuous.

A somewhat different use of composite estimate occurs when a real-valued function $f(x)$, defined by $a \leq x \leq b$, is to be bounded from above by a possibly low constant; we may be able to develop two different upper estimates, $L(x)$ and $U(x)$. The lower-end estimate $L(x)$ bounds $f(x) : f(x) \leq L(x)$ for $a \leq x \leq b$, and it is good near $x = a$ but deteriorates as x approaches b; similarly, $f(x) \leq U(x)$ for $a \leq x \leq b$, the upper-end estimate $U(x)$ is good near $x = b$ but not near $x = a$. If $U(x)$ is monotone decreasing, $L(x)$ monotone increasing, and both are continuous, we find

$$f(x) \leq \min_x (L(x), U(x))$$

and therefore the best constant bound is $f(x) \leq K$ where K is the unique root of the equation $L(x) = U(x)$.

As an example of this type of double estimation we shall consider the two-dimensional case of the Borsuk conjecture. The general n-dimensional case runs as follows: a subset X of the n-dimensional Euclidean space E^n, of diameter ≤ 1, is a union of $n + 1$ sets of diameter < 1. It is easy to show that $n + 1$ is the minimal number possible. A proof of the special case of the Borsuk conjecture when X is sufficiently smooth is given in the section on convexity of Chapter 1. The Borsuk conjecture has been proved in two and three dimensions but for no higher n. Here we give a proof of the best possible result in the plane: let X be a plane set of diameter ≤ 1 (i.e., with no pair of points further than one unit apart); then X is a union of three sets of diameter $\leq 3^{1/2}/2$, and the constant is best possible.

We start with our arbitrary X of diameter ≤ 1 and we perform three reductions of the problem. First, it may be assumed without loss of generality that X is closed for a bounded set and its closure has the same diameter. Next, it may be similarly assumed that X is convex for a bounded set and its convex hull has the same diameter. Finally, it may be assumed that X is adjunction-complete, i.e., that its diameter actually increases when any new points are added to it. Now, it is well known, and rather easy to prove, that a set which is adjunction-complete is of constant width, and vice versa. (Here a plane set X is of constant width a if for any direction u in the plane the two lines which support X from opposite sides and are perpendicular to u, are the fixed distance a apart). We give a proof which applies in any dimension. To show that a set X of constant width is complete we consider a point $p \notin X$ and we let q be the point in the boundary ∂X of X, which minimizes px, $x \in X$. It follows that the plane Q through q orthogonal to pq supports X at q; let Q_1 be the plane parallel to Q and supporting X from the other side, at q_1 say. It follows then that qq_1 is the diameter D of X and hence $q_1 p > D$.

Conversely, let X be complete, of diameter D, and suppose it is not of constant width. Let d be the minimal width of X, which is the shortest distance between a pair of parallel support planes, enclosing X between them. Let these planes be P_1 and P_2 supporting X at p_1 and p_2; the minimal width d implies that $p_1 p_2$ is orthogonal to P_1 and P_2 and of length d. By the hypothesis of completeness there is a point $q \in X$ such that $p_1 q = D$. Further, since $p_2 \in X$ and $q \in X$, X contains also, by completeness, the cigar-shaped convex body $X(p_2, q)$ defined to be the intersection of all balls of radius D, which pass through p_2 and q. Suppose now that $d < D$, then $X(p_2, q)$, and therefore X, lies on both sides of P_2, which is a contradiction.

Therefore, to prove our original assertion it suffices to show that a closed convex set X of constant width 1 is partitionable into three sets of diameter $\leq 3^{1/2}/2$ but no better.

Let X be such a set; we show that an equilateral triangle can be inscribed into X in arbitrary orientation. With reference to Figure 1 we consider a small equilateral triangle two of whose vertices are on the boundary of X and the side joining them is parallel to a given line L. When the triangle is small enough it lies inside X, in particular, the third vertex v_1 is in the interior of X. When the triangle increases, the third vertex eventually arrives at v_2 outside

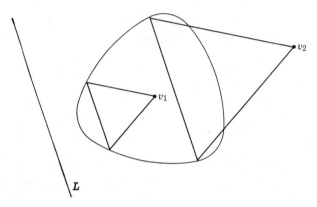

FIGURE 1.   Equilateral triangles and constant width.

X; since it moves continuously it must cut the boundary X somewhere, and we have our required inscribed equilateral triangle.

We have now some useful structure in our problem; in particular, we have the quantity $a(X)$ to work with—the side of the largest equilateral triangle inscriptible into X, and we perform our double estimate with respect to it. If $a(X)$ is small we cut up X into three parts shown in Figure 2a, if $a(X)$ is big we cut it up as in Figure 2b. Let $B(a)$ be the infimum (actually a minimum but

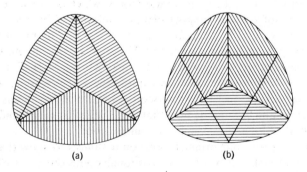

(a)   (b)

FIGURE 2.   Partitioning a set of constant width.

that is irrelevant here) of all numbers w such that X, with $a = a(X)$ fixed, can be decomposed into three parts of diameter $\leq w$. Then the lower-end and upper-end estimates of Figure 2a and 2b can be shown to be

$$L(a) = a, \qquad U(a) = 3^{1/2} - a$$

so that

$$B(a) \leq \min(a, 3^{1/2} - a) = 3^{1/2}/2$$

which was the assertion. The example of the circular disk shows that the constant is best possible.

8. SEQUENTIAL LOCALIZATION AND MAXIMA OF UNIMODAL SURFACES

Let $f(x)$ be a continuous strictly decreasing function defined for $0 \leq x \leq 1$ and suppose that $f(1) < 0 < f(0)$. The unique zero z of $f(x)$ may be localized by a sequential testing procedure: we test the sign of $f(1/2)$, if this is plus we test next the sign of $f(3/4)$, if minus the sign of $f(1/4)$, and so on. In this way n tests confine z to an interval of length 2^{-n}; since

$$\max_{x}[\min(x, 1 - x)] = 1/2$$

we can do no better than 2^{-n}. This numerical procedure, though crude and slow, has several features to recommend it: simplicity, stability against computational noise, etc. It might be actually used to determine the zero of such a function as

$$f(x) = \int_{-\infty}^{\infty} F(x, y)g(y)\,dy$$

where F and g are fairly complicated. Similarly, let $f(x)$ be a function defined for $0 \leq x \leq 1$ and having exactly one maximum, say at m. We can localize m by choosing x_1 and x_2 such that $0 < x_1 < x_2 < 1$, and testing the sign of $f(x_2) - f(x_1)$. If this is positive, m cannot be in the interval $[0, x]$ for there is only one maximum; if the difference $f(x_2) - f(x_1)$ is negative then m cannot be in $[x_2, 1]$. By repeating the procedure we can localize the maximum m to an arbitrarily small interval. How small an interval can be achieved in n steps? Here the optimality is somewhat harder to investigate than before.

We dualize the problem by asking: What is the *longest* interval L_n on which a unimodal ($=$ having one maximum) function may be given so that in n steps the maximum can be localized to a unit interval? Once L_n is known, then $1/L_n$ is the shortest interval to which the maximum m can be localized in n steps starting with the interval $[0, 1]$. Let us choose two testing points x_1 and x_2, with $0 < x_1 < x_2 < L_n$. We test the sign of $f(x_2) - f(x_1)$; if it is positive then $m \in [x_1, L_n]$ whose length is $L_n - x_1$, and one test-point, x_1, has been used and will be of no further use to us on the interval $[x_1, L_n]$—the other point x_2 and the value $f(x_2)$ will still be of use. Therefore $L_n - x_1 \leq L_{n-1}$. It is also possible that the maximum m might be on $[0, x_1]$; in this case two test-points have been used up and so $x_1 < L_{n-2}$. Therefore

$$L_n \leq L_{n-1} + L_{n-2};$$

we consider now the Fibonacci sequence $F_0, F_1, F_2, \ldots$ defined by $F_0 = F_1 = 1$,

$$F_n = F_{n-1} + F_{n-2}, \qquad n = 2, 3, \ldots$$

and we have $L_n < F_n$; we show that for any $\varepsilon > 0$ we can achieve $L_n > F_n - \varepsilon$. Retroactively, we remark here that the maxima and minima should be replaced in the preceding by suprema and infima.

We formulate now another problem: let $f(x, y)$ be an (arbitrarily smooth) function defined on the unit square

$$S = \{(x, y): 0 \leq x \leq 1, 0 \leq y \leq 1\}$$

and having there exactly one maximum, say, at (a, b). Then no sequential procedure of the type we have used will localize the position (a, b) of the maximum. We prove it in a rather strong sense, allowing not only just sign comparisons as before, but also any use of function values. Suppose that $p_i = (x_i, y_i)$, $i = 1, \ldots, n$, are any n points in S, and let f_i, $i = 1, \ldots, n$, be the values $f(x, y)$ takes at those points; then there exists a function $F(x, y)$ defined on S, arbitrarily smooth, taking up the same values f_i at each point p_i, but having exactly one maximum which can be made to lie in an arbitrarily small neighborhood of *any* point of S. Thus the n values of f are of no use whatever in localizing the maximum m.

On the basis of those examples for the interval and for the square, the reader may ask himself what precisely is behind our ability to localize in one case, and inability in the other case. In both cases the position of m can be *restricted* to a proper subset, but whereas for an interval every *restriction* is a *contraction*, this is not true for the square. Thus, what is really behind localizability, is our ability or inability to confine m to a subset of smaller diameter. On the other hand, the reader may wish to show that the maximum m of the function $f(x, y)$ becomes localizable as soon as we place a bound on the growth of $f(x, y)$, such as

$$\left|\frac{\partial f}{\partial x}\right| \leq M \quad \text{and} \quad \left|\frac{\partial f}{\partial y}\right| \leq M$$

or the Lipschitz condition

$$|f(x_2, y_2) - f(x_1, y_1)| \leq M[(x_2 - x_1)^2 + (y_2 - y_1)^2]^{1/2}.$$

The reason for this is precisely that now we can confine m by a suitable test from the unit square to a set of smaller diameter.

We consider one further question connected with sequential localization on an interval. Even in the relatively simple case of localizing a single maximum on $[0, 1]$ it was not entirely simple or obvious what the optimal procedure was. For some more complicated problems the optimal procedure may be quite difficult to find; indeed, in some cases it is as yet unknown, for instance, for localizing the unique zero of the kth derivative of $f(x)$ on $[0, 1]$. This suggests the idea of random localization. We indicate this on the example of localizing the unique zero m of a continuous steadily decreasing function $f(x)$, with $f(0) > 0 > f(1)$. The optimum procedure was to test the sign of $f(1/2)$, then to test the sign of $f(3/4)$ (if $f(1/2) > 0$) or the sign of $f(1/4)$ (if $f(1/2) < 0$), and so on. Suppose now that we choose the number x_1 uniformly at random on $[0, 1]$ and test the sign of $f(x_1)$; if $f(x_1) > 0$ then we choose x_2 uniformly at random on $[x_1, 1]$ and test the sign of $f(x_2)$. If $f(x_1) < 0$ we do the same but with x_2 uniformly at random on $[0, x_1]$. Let e_n be the expected value of the length of the interval to which m has been confined after n applications of the random method. Then the reader may wish to show that for large n

$$e_n \simeq \frac{8}{\pi} \sqrt{m(1 - m)}(2/3)^n.$$

9. WORKING BACKWARD FROM THE END AND WORKING FROM BOTH ENDS

Sometimes a problem may be solved so to say backward, by an analysis of the answer. Suppose, for instance, that we wish to prove the following: given any nine distinct points of the integer lattice in E^3, the straight segment determined by some two of the nine points contains another lattice point (which is not necessarily one of the nine given). We analyze the "9"; it could be 3^2 (it isn't) or $3 + 6$ (it isn't) or $2^3 + 1$ (which is it). Why $2^3 + 1$? An integer lattice point (i, j, k) has integer coordinates i, j, k; the three positions, with two possibilities, even or odd, for each position, give us 2^3 parity possibilities in all. Add 1 more and it follows that of the nine points p_i, some two, say p_1 and p_2, must have all three coordinates of the same parity. Hence $(p_1 + p_2)/2$ is also a lattice point, q.e.d.

A less obvious example is the Cayley formula for the number c_n of labeled trees on n vertices, considered in Example 5 of Section 10, Chapter 3, $c_n = n^{n-2}$. By analyzing this we are led to the possibility of a 1 : 1 correspondence between the distinct labeled trees on n vertices and ordered $(n - 2)$-tuples $(a_1, a_2, \ldots, a_{n-2})$, where each a_i is an integer and $1 \le a_i \le n$. The idea of such a correspondence may be obtained from the usual proof that in any tree the number of vertices exceeds the number of edges by 1. This proof goes as follows. Consider a tree, for instance, the tree of Figure 1a. Let us consider any vertex of valency 1, i.e., with just one edge emanating from it; for instance, we can take the vertex 3. We erase both the vertex 3 and the edge 32, obtaining the tree of Figure 1b. We repeat the same operation, obtaining eventually the tree of Figure 1c. Next, we notice that any removal operation preserves the difference $D = $ (no. of edges) $-$ (no. of vertices). Since for the tree of Figure 1c $D = 1$, it follows that $D = 1$ always, q.e.d.

For a tree with n vertices $n - 2$ removals would be necessary; this suggests a connection with our ordered $(n - 2)$-tuple $(a_1, a_2, \ldots, a_{n-2})$. Our first removal concerned the vertices 3 and 2, but we could have equally well removed the vertex 6 and the edge 61 to begin with. However, to keep order, we shall always remove the lowest vertex possible, thus vertex 3 was the right one. For a_1 we take not 3, the vertex removed, but 2, the vertex from which vertex 3 was removed. Now the $(n - 2)$-tuple for our case is uniquely determined: (2, 2, 2, 1, 1). Conversely, given the ordered 5-tuple (2, 2, 2, 1, 1), can we reconstitute from it the whole tree of Figure 1a? Yes, for the first three vertices removed must have been 3, 4, 5 since they could not be 1 or 2, and these three have been all connected to 2 as is clear from the 5-tuple. This gives us the fragment of Figure 1d. The fourth vertex to be removed was connected to 1, and that vertex was 2. The final vertex to be removed was also connected to 1, and it was 6. The tree is now completely reconstituted. Since this idea carries over to the general case, we have our 1 : 1 correspondence, and so an independent proof of Cayley's formula $c_n = n^{n-2}$, due to Prüfer [60]. As observed by Moon [53], this proof yields rather more than the proof of Cayley's formula. For, let the vertex i have valency d_i; using the principle of counting the same thing in two different ways, we count the number of edges once as $\sum_1^n d_i$, and once as $2(n - 1)$; in this counting each edge was counted exactly twice, once from each end. Thus we must have

$$\sum_1^n d_i = 2(n - 1). \tag{1}$$

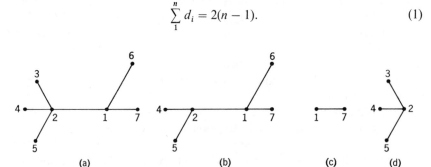

(a)   (b)   (c)   (d)

FIGURE 1. A small tree.

Next, as Moon observes, Prüfer's proof extends to the following: if (1) holds, then the number of trees in which vertex i has valency d_i, is the multinomial coefficient

$$\binom{n-2}{d_1 - 1 \quad d_2 - 1 \quad \cdots \quad d_n - 1}. \tag{2}$$

Conversely, it follows now that

$$c_n = \sum_{d_1, d_2, \ldots, d_n} \binom{n-2}{d_1 - 1 \quad d_2 - 1 \quad \cdots \quad d_n - 1}$$

so that $c_n = n^{n-2}$ again, since the right-hand side is by the multinomial theorem equal to

$$(1 + 1 + \cdots + 1)^{n-2} \qquad (n 1\text{'s}).$$

We consider next some problems which may be solved by working from both ends. The analogy here is to the drilling of a tunnel; since the cost of drilling increases faster than the first power of the length, it pays to drill from both ends, provided that we can make the ends meet. For instance, in proving an identity $f(x) \equiv g(x)$ we may be able to show that $f(x) \equiv f_1(x)$ and $g(x) \equiv g_1(x)$, then $f_1(x) \equiv f_2(x)$ and so on, and each successive f_i is "more like" the g's than the preceding one. As a less trivial example of working from both ends we consider the following problem: given a connected graph G with two distinguished vertices a and z, find the path from a to z which consists of the smallest number of edges. For instance, G may represent a system of roads and a and z two cities, there is a unit toll-payment for crossing each edge and we ask for the lowest total toll trip from a to z.

We start from the vertex a marking off all its immediate neighbors, and we do the same for z; next, we mark off the (as yet unmarked) nearest neighbors of the nearest neighbors, and so on. As soon as a vertex has been reached from both ends we have our desired minimal path. A simple modification takes care of the case when several vertices might be reached from both ends simultaneously, and gives us all the minimal paths (if there are several).

This problem could also be solved by working from the a-end only; possible advantages of working from both ends are: (a) smaller number of operations; (b) if the problem is being solved by a computing machine, we may arrange for a more highly parallel course of computation, thus cutting down on time.

Similar technique will occasionally help in other graph minimization problems. The reader may observe some similarities to certain problems of Section 3 of this chapter. These similarities depend on tangency properties of circles and ellipses, which carry over to metrics other than the ordinary Euclidean distance, and which apply to discrete as well as continuous cases. A simple case in the plane was concerned with Steiner's problem; it is illustrated in Figure 2. The point S which minimizes the sum $|AS| + |BS| + |CS|$ is such that the ellipse through S, with A and B as foci, is tangent at S to the circle through S with C as center. Hence $\not\prec ASC = \not\prec BSC$, and by symmetry $\not\prec ASC = \not\prec BSC = \not\prec BSA = 120°$. Going over to the discrete case the reader may wish to devise an algorithm for solving the following problem: given a connected graph G and three distinguished vertices a, b, c of G, to find the vertex z (or vertices z) of G, for which the sum of distances $|az| + |bz| + |cz|$ is minimized. Here the distance $|xy|$ between two vertices is the smallest number of edges forming a path from x to y.

10. COMPUTING π

The history of computing π ($= 3.14159\ 26535\ 89793\ 23846\ 26433\ 83279\ 5028$ ...) starts with the value 3, implicit in a sentence from the Bible referring to a round washbasin built at Solomon's order about the year 1000 B.C., and continues to a recent computation of approximately half a million digits on an electronic calculator. The standard approximating fraction is $22/7$; this is related to the work of Archimedes who used inscribed and circumscribed regular polygons to show that $3\ 10/71 < \pi < 3\ 1/7$. However, $22/7 = 3.143 ...$ whereas a far superior, and easily memorizable, fraction is $355/113 = 3.1415929 ...$, due to a medieval Chinese mathematician. A value $\pi = 3.141552$ was given between 125 A.D. and 151 A.D. by Claudius Ptolemaeus of Alexandria, the author of the Ptolemaic system of astronomy. The approximation $\pi = \sqrt{10}$ is due to Brahmagupta, a fifth-century Hindu mathematician.

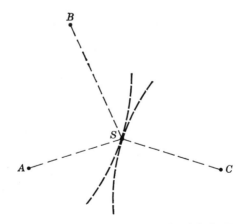

FIGURE 2.   Tangency and graph minimization.

The greatest of the early calculators of π was the Dutch mathematician Ludolph van Ceulen (1540–1610), a professor of military engineering at the University of Leyden, who, together with his wife, Adriana Symonsz, calculated π to 35 decimal places. This fact, together with the full decimal development, was stated on Ludolph's tombstone in Leyden. Due to his work, π was occasionally called in older European literature Ludolph's number.

The first exact formula for π, which also happens to be the first infinite product ever used, was given by Francois Vieta (1540–1603):

$$\frac{2}{\pi} = \frac{\sqrt{2}}{2} \frac{\sqrt{2 + \sqrt{2}}}{2} \frac{\sqrt{2 + \sqrt{2 + \sqrt{2}}}}{2} \cdots ;$$

it is proved in Chapter 3. Vieta, like his countryman Pierre Fermat, a lawyer and parliamentary councillor, was the author of many books on mathematics which he had printed at his own cost and sent to French and foreign scientists. He is the originator of much of the modern arithmetic and algebraic notation.

The expressions

$$\arctan x = x - x^3/3 + x^5/5 - \cdots \tag{1}$$

and in particular, for $x = 1$,

$$\pi/4 = 1 - 1/3 + 1/5 - \cdots \tag{2}$$

are due to Leibniz (1646–1716). Two mathematicians whose names are associated with π, are Lord Brouncker (1620–1684) and his friend John Wallis (1616–1703), who were, respectively, the chancellor and the chaplain to Charles the Second of England. Brouncker was one of the founders and the first president of the Royal Society. Wallis's contribution is the infinite product

$$\frac{\pi}{2} = \prod_{n=1}^{\infty} \frac{(2n)(2n)}{(2n - 1)(2n + 1)}.$$

The continued fraction

$$\frac{4}{\pi} = 1 + \cfrac{1^2}{2 + \cfrac{3^2}{2 + \cfrac{5^2}{2 + \cdots}}}$$

Convergence is slow.

is due to Brouncker. It was probably obtained by applying a special case of a transformation which was later developed by Euler: if

$$s = x/a - x^2/ab + x^3/abc - \cdots$$

then

$$x/s = a + \cfrac{ax}{b - x + \cfrac{bx}{c - x + \cfrac{cx}{d - x + \cdots}}}$$

The irrationality of π was proved in 1761 by J. H. Lambert (the proof is given in Chapter 6). Finally, the transcendency of π was proved in 1882 by F. Lindemann, who followed closely the earlier proof of the transcendency of e due to C. Hermite (1873).

Starting with (1) and (2), one obtains a variety of arc-tan type formulas for π:

$$\pi/4 = \text{arc tan } \frac{1}{2} + \text{arc tan } \frac{1}{3}$$

$$= 2 \text{ arc tan } \frac{1}{3} + \text{arc tan } \frac{1}{7}$$

$$= 2 \text{ arc tan } \frac{1}{5} + \text{arc tan } \frac{1}{7} + 2 \text{ arc tan } \frac{1}{8}$$

$$= 8 \text{ arc tan } \frac{1}{10} - \text{arc tan } \frac{1}{239} - 4 \text{ arc tan } \frac{1}{515}$$

$$= 5 \text{ arc tan } \frac{1}{7} + 2 \text{ arc tan } \frac{3}{79}$$

$$= 4 \text{ arc tan } \frac{1}{5} - \text{arc tan } \frac{1}{239}, \text{ etc.} \qquad (3)$$

Many such formulas can be created by using the identity attributed in [10] to C. L. Dodgson (who, as Lewis Carroll, wrote "Alice in Wonderland" and "Through the Looking Glass"): if $qr = 1 + p^2$ then

$$\text{arc tan } \frac{1}{p} = \text{arc tan } \frac{1}{p + q} + \text{arc tan } \frac{1}{p + r}.$$

However, this identity was already given by Euler in 1737. As shown in Section 9, Chapter 3, we have the expansion

$$(1 - y^2)^{-1/2} \text{ arc sin } y = y + \frac{2}{3} y^3 + \frac{2 \cdot 4}{3 \cdot 5} y^5 + \cdots$$

and consequently

$$\arc\tan\frac{m}{n} = \frac{mn}{m^2 + n^2}\left[1 + \frac{2}{3}\frac{m^2}{m^2+n^2} + \frac{2\cdot 4}{3\cdot 5}\left(\frac{m^2}{m^2+n^2}\right)^2 + \cdots\right];$$

if this is applied to expanding the arc-tan terms in (3), we find that considerable simplification (numerically) results if $m^2 + n^2$ is either a power of 10 or a suitable divisor of such a power. For instance, using the penultimate formula of (3) and noting that

$$2(1^2 + 7^2) = 10^2, \qquad \overset{16}{\cancel{12}}(3^2 + 79^2) = 10^5$$

we have

$$\frac{\pi}{4} = \frac{7}{10}\left[1 + \frac{2}{3}\left(\frac{2}{100}\right) + \frac{2\cdot 4}{3\cdot 5}\left(\frac{2}{100}\right)^2 + \cdots\right]$$

$$+ \frac{7584}{100000}\left[1 + \frac{2}{3}\left(\frac{144}{100000}\right) + \frac{2\cdot 4}{3\cdot 5}\left(\frac{144}{100000}\right)^2 + \cdots\right]$$

with the resulting savings on the number of divisions.

However, strange as it may appear, it seems that no really fast way of computing π is known. If E_n is the truncation error resulting from stopping after n terms in a series, or at the nth stage of some iteration process, then we call the numerical procedure exponentially fast if for large n, $E_n \sim c^n$ with some constant c, $0 < c < 1$. On the other hand, the iterative procedures for the arithmetic-geometric mean of Gauss or the Newton–Raphson algorithm (of Chapter 2) are hyperexponentially fast: here $E_n \sim c^{n^a}$ with $0 < c < 1$ and $a > 1$. No such hyperexponentially fast procedure appears to be known for computing π.

We finish this section with two formulas which allow us to compute π and $e^{-\pi}$ by means of very fast convergent series. The genesis of our formulas is in the theory of elliptic functions and we begin by assembling some necessary apparatus, [77]. The integral relation

$$x = \int_0^y \frac{du}{\sqrt{(1 - u^2)(1 - k^2 u^2)}}, \qquad 0 \le k \le 1$$

may be inverted, to yield y as a function of x. In analogy to

$$x = \int_0^y \frac{du}{\sqrt{1 - u^2}} \;(= \arc\sin y) \to y = \sin x,$$

here we get

$$y = \operatorname{sn}(x, k)$$

which is the Jacobian elliptic function sn, with the modulus k. The complementary modulus k_1 is given by

$$k^2 + k_1^2 = 1, \qquad 0 \le k_1 \le 1.$$

Let K and K_1 be the complete elliptic integrals

$$K = \int_0^1 \frac{du}{\sqrt{(1 - u^2)(1 - k^2 u^2)}} = \int_0^{\pi/2} \frac{d\phi}{\sqrt{1 - k^2 \sin^2 \phi}},$$

$$K_1 = \int_0^1 \frac{du}{\sqrt{(1 - u^2)(1 - k_1^2 u^2)}} = \int_0^{\pi/2} \frac{d\phi}{\sqrt{1 - k_1^2 \sin^2 \phi}},$$

then it turns out that $\operatorname{sn}(x, k)$ is doubly periodic:

$$\operatorname{sn}(x + 4K, k) = \operatorname{sn}(x + 2iK_1, k) = \operatorname{sn}(x, k).$$

From the theory of the elliptic theta-functions [77], we get the following relation: if

$$b = \frac{1}{2} \frac{1 - \sqrt{k_1}}{1 + \sqrt{k_1}} \qquad \text{and} \qquad q = e^{-\pi K/K_1}$$

then

$$(1 + 2q^4 + 2q^{16} + 2q^{36} + \cdots)b = q + q^9 + q^{25} + q^{49} + \cdots. \tag{4}$$

Substituting a McLaurin series

$$q = \sum_{n=0}^{\infty} a_n b^n$$

into (4) and comparing the coefficients, we find

$$q = b + 2b^5 + 15b^9 + 150b^{13} + 1707b^{17} + \cdots. \tag{5}$$

Next, we form the series

$$q/b = 1 + 2b^4 + 15b^8 + 150b^{12} + 1707b^{16} + \cdots = 1 + X,$$

say. Taking logarithms and using the power series for $\log(1 + X)$, we find

$$\log q = \log b + 2b^4 + 13b^8 + \frac{368}{3} b^{12} + \frac{2701}{2} b^{16} + \cdots. \tag{6}$$

Suppose now that $k = k_1 = 1/\sqrt{2}$ so that $K = K_1$, $q = e^{-\pi}$, and

$$b = \frac{1}{2} \frac{\sqrt[4]{2} - 1}{\sqrt[4]{2} + 1}. \tag{7}$$

Then (5) and (6) give us the following series for $e^{-\pi}$ and for π:

$$e^{-\pi} = b + 2b^5 + 15b^9 + 150b^{13} + 1707b^{17} + \cdots \qquad (8)$$

$$\pi = \log\frac{1}{b} - 2b^4 - 13b^8 - \frac{368}{3}b^{12} - \frac{2701}{2}b^{16} - \cdots ; \qquad (9)$$

since the value of b, as given by (7), is a small number

$$b = 0.0432136168629448960219378 \ldots$$

our series have terms which fall off very rapidly. The first term alone, $\log 1/b$, already gives π correctly to better than five decimals. For comparison purposes we give below the values of π and of the first five partial sums of the series in (9), each partial sum being calculated to the first digit on which it disagrees with π:

$$\pi = 3.14159\,26535\,89793\,23846\,26433$$
$$3.14159\,9$$
$$3.14159\,26537$$
$$3.14159\,26535\,89798$$
$$3.14159\,26535\,89793\,2386$$
$$3.14159\,26535\,89793\,23846\,265.$$

By (8) $e^{-\pi}$ is approximately equal to b, to better than six digits' accuracy. In view of (7) this may also be expressed thus: $e^{-\pi}$ very nearly satisfies the equation

$$16x^4 - 96x^3 + 24x^2 - 24x + 1 = 0.$$

11. THE GENERALIZED FIRING SQUAD SYNCHRONIZATION AND THE LABYRINTH PROBLEMS

The subjects of this section belong properly to the theories of automata and of pattern perception. They are included here because, in addition to considerable intrinsic interest, they have connections with what might be called computational geometry and, more particularly, computational topology. This last point is brought out especially in the "knot conjecture" made later on.

We state first the original firing squad synchronization problem as it arose in connection with turning on all parts of a self-reproducing automaton at once [54]. The "firing squad" is a finite though arbitrarily long chain of n equispaced machines, called soldiers, standing in a line. The soldiers are identical finite-state synchronous automata: the state of each soldier at the time $t + 1$ depends on its state and the state of two immediate neighbors (or one such, for the two extreme soldiers) at the time t. The leftmost soldier is called a general, the rightmost a sergeant. At the time $t = 0$ the general gives the order "fire when ready", all the other soldiers are inactive then. The object is to construct the soldiers so that they all "fire" together (are synchronized). The difficulty is that one kind of soldier, fixed in advance, has to suffice for every size n of the array. We can say, roughly, that the soldiers operate by clock, can be trained to follow infallibly any strictly prescribed system of instructions but no others, and can communicate only with their immediate neighbors. Thus, they can be given the ability to count to 2, 10, or 10^{10} but not to count to (unspecified) n.

With suitable encoding, signals can be propagated up and down the line, with top speed 1 (one spacing per one time unit); delays can be arranged so that the signals propagate at lower speeds. In a graphical representation, with the soldier line on one axis and time on the other, we have some slight similarity with the world lines and cones of relativity theory. The general starts two signals, the first with the speed 1, the second with the speed 1/3. The fast signal is reflected back by the sergeant and the two signals cross in the middle of the array. Paying some attention to the parity of n, we promote the middle soldier or the middle two soldiers to general. New fast and slow signals are emitted from the new general or new generals into the two halves of the array, and the process is repeated on and on. As soon as everybody and his neighbors are generals, the firing occurs.

What has been outlined is only one, though perhaps the simplest, way of solving the problem. The synchronization takes about $3n$ time units and the number of states of each soldier depends on coding ingenuity but 20–30 is a reasonable number. It can be shown that the firing cannot occur at a time $<2n - 2$ which is the minimum time for a signal to go forth and return to the original general. Actual firing time $2n - 2$ is achievable but at the cost of greatly complicated soldiery. It is apparently not known yet what is the minimum number of states of a soldier which achieves synchronization.

Before generalizing the problem we restate it slightly. First, we democratize the army by doing away with the general and the sergeant and we pose the problem as follows. Given a row of n identical soldiers as before, a soldier is chosen at random and put into the "fire when ready" state at $t = 0$, the synchronization is to occur just as before (though its time will depend on which soldier was chosen initially). We solve this problem by reducing it to the previous one: the initial soldier starts two pulses, one to the left and one to the right, we re-introduce ranks by promoting the leftmost machine to general and the rightmost one to sergeant, and now we have exactly the previous situation.

Next, we generalize the whole problem to arbitrary dimension. In the d-dimensional Euclidean space we consider the set L of all points of the integer lattice; these are the points $(x_1, \ldots, x_d)$ with every x_i an integer. A finite subset S of L is called a shape if it is connected, i.e., if every two points of S can communicate in S via a chain of immediate neighbors (points a unit distance apart). A pattern is any infinite collection of shapes (some further restrictions may be added here, for instance, one may require that if S is in the pattern so is any shape similar or congruent to S). A pattern P is weakly distinguishable (or synchronizable) if there is a finite-state synchronous machine M, such that the following happens. Let any shape S be chosen from P, let a copy of M be placed at every point of S, let any copy be chosen at random and put into the "fire when ready" state; we now require that the whole array of machines M sitting at the points of S, should synchronize just as in the one-dimensional problem. Here we suppose that the machines can send signals as before, have a sense of direction ($=$can distinguish between right, left, up, down, etc.), and also have a sense of immediate environment ($=$can tell whether an immediate neighbor exists in any one of the $2d$ possible directions). The time of synchronization will in general depend on the size of S as well as on which machine in S was initially chosen. A pattern P is strongly distinguishable, or strongly synchronizable, or briefly, distinguishable, if the synchronization just described will take place for any shape in P but for no others. Here, however, it might happen that the particular machine M, admissible for a pattern P, is such that shapes not in P will also be synchronized provided that suitable (though not arbitrary) initial points will be chosen for the "fire when ready" state.

As an example we take $d = 2$ and we define a shape to be any rectangular array of lattice points in the plane; the set of all such rectangular shapes is the pattern R, the rectangle. Or, we can do the same with square arrays only, obtaining the pattern Q, the square. It is not hard to show that R is distinguishable by a suitable machine M. We obtain M by putting together two copies, M_1 and M_2, of the machine which synchronizes one-dimensional arrays. The first copy M_1 is called the horizontal component of M, M_2 is the vertical component. When any M is chosen and put into the "fire when ready" state, its horizontal component starts synchronizing all the horizontal components of machines with the same y-coordinate as the one originally chosen. The "fire" command is replaced by the "fire when ready" command to all the vertical components, and now the synchronization proceeds vertically. The reader may wish to verify that R is strongly distinguishable by the machine M just described; also, a suitable modification of M will strongly synchronize the pattern Q, the square.

With this generalized setting we see the one-dimensional firing squad synchronization problem in its proper light: a pattern recognition problem for one-dimensional arrays. Now, in one dimension there is only one possible pattern—the connectivity—since nonconnected shapes are not admissible. However, in higher dimensions we have lots of patterns and of problems about patterns. It may be remarked that with this approach to pattern theory we have automatically a good complexity measure with which to compare different patterns, namely, the number of states of simplest machine which suffices for synchronization. For instance, the reader may wish to show in this way that the pattern R, the rectangle, is simpler than the pattern Q, the square. In addition to various questions of minimal firing time and minimal complexity we have the important basic question, which is perhaps the most direct generalization of the original one-dimensional problem: Is the class of all connected d-dimensional shapes distinguishable? The reader may wish to show that the answer is yes.

There is one rather natural question: Is there a nondistinguishable pattern? No good explicit bona fide example seems to be known. We formulate now a conjecture which submits a likely and rather interesting candidate. We take $d = 3$ and we call a connected finite set S of points in the three-dimensional integral lattice a closed curve if every point in S has exactly two nearest neighbors in S. The set of all closed curves S is a pattern, say A. A slight modification of the one-dimensional synchronization problem (bending the linear array and joining the ends to form a loop, disconnecting a loop to form an essentially linear array) shows that A is strongly distinguishable. Since every curve in A is topologically either knotted or unknotted, A splits into two complementary disjoint patterns K and U. We conjecture that neither K nor U is distinguishable.

A much more general conjecture might be attempted concerning the distinguishable vs. indistinguishable patterns. For illustration we let the dimension be 3 and we consider several three-dimensional patterns: (a) R, the parallelopipeds (i.e., the three-dimensional equivalent of rectangles), (b) Q, the cubes, (c) C, all connected three-dimensional shapes, (d) K, closed curves which are knots. It may be observed that in some sense these four patterns are progressively more "global." The most "local" pattern is R, and we can decide for any candidate A whether it belongs to R by examining it piece by piece. That is, we choose in turn each point of A and look at some restricted neighborhood of the chosen point, for instance, at all points whose distance from the chosen one is $\leq 3^{1/2}$. Moreover, in passing from one neighborhood to another we manage to forget all that we saw before.

No such technique will do for the pattern Q. However, a set A belongs to Q if and only if it belongs to R and also satisfies a certain small number of additional conditions. When we come to C we can convince ourselves that no bounded number of simple additional conditions will do. Continuing in this way, we are led to conjecture that there may be some measure of "globality" which, if too big, makes a pattern nondistinguishable.

Finally, it may be observed that the restriction to shapes which are parts of integer lattices is not really essential. We can frame our problems for classes of finite connected graphs, making some provision for lack of orientation (now there is no right, left, up, down, etc.). We obtain further problems because we may ask now, with proper formulation, whether trees, complete graphs, k-partite graphs, etc., are distinguishable.

The transition to graphs suggests also another class of problems. We suppose that a finite connected graph represents a labyrinth as shown in Figure 1, the edges marked by small letters are corridors, and the vertices marked by capital letters are the junctions. There may be blind corridors (like s, dead-end G counts as a junction), over- and under-passes, corridors which start from a junction and lead back to it, and pairs of junctions connected by more than one corridor. One of the junctions is marked "entrance" and another is marked M (for Minotaur). A machine T (for Theseus) is stationed at the entrance; it can move at will along the corridors of the labyrinth and it recognizes a junction whenever it comes to one. Further, it can count to arbitrarily high limit and it has arbitrarily capacious memory. However, it is not allowed to look ahead or do any marking of corridors or junctions visited, and it cannot recognize them once they have been visited. Thus Theseus is not a finite-state machine like the previous soldiers. Finally, we suppose that Theseus recognizes M (or vice versa). Now the object is to program Theseus so that it enters the labyrinth, finds M, and returns back to the entrance (other tasks might be assigned such as finding the *shortest path* to M, or storing in a suitable code the map of the entire labyrinth).

We start Theseus in the doorway facing the entrance A. Starting with the right-hand corridor b, T counts all the corridors emanating from A; there are just two and so b is 1 and a is 2. Suppose T picks corridor 2 (that is a) and

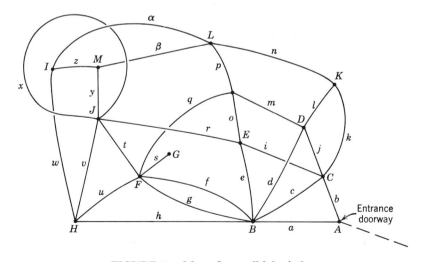

FIGURE 1.   Map of a small labyrinth.

comes along to the junction B. Standing in the corridor a and facing B, T counts all the corridors from B just as before, now c is 1, d is 2, e is 3, f is 4, g is 5, h is 6, and a is 7. Let us say that T chooses the fourth corridor, f, and walks toward F. The same procedure is repeated and T picks the first corrridor, s, getting into the blind junction G. Here T takes the only available way out, s, getting back to F (without, of course, recognizing F as having been visited). From F, standing in s, T picks the fifth corridor, f, getting him to B, whence the second corridor takes him to H. Theseus's travels are shown in the following table:

at the junction	took corridor number	which is
A	2	a
B	4	f
F	1	s
G	1	s
F	5	f
B	2	h
H		

We suppose that Theseus memorizes the numbers 2, 4, 1, 1, 5, 2 in this order. Suppose that now it decides to return back to the entrance. Starting from its position in h, T turns about and moves back to the first junction which is B. Now it chooses the second corridor counting *from the left*, f, moves along it to F, chooses the fifth corridor from the left, s, and so on; it executes the trip coded by the reverse sequence: 251142, each time counting corridors from the left. Eventually Theseus returns to the door A, having executed the trip coded by the number 251142 of length 6. Thus Theseus can always return to A after performing any trip. It therefore executes all the trips of length 1, then all the trips of length 2, 3, 4, ..., each time returning to A. Eventually T reaches M, and then returns to A.

Suppose now that T is a finite-state machine, but that it has some means of marking off a corridor each time that corridor is crossed. The reader may wish to show that a suitably programmed T will reach M (and return) never passing through a corridor more than twice.

5

INTEGRALS

The elementary techniques of integration by parts, substitution, and partial fractions will be taken for granted. In this chapter we will be concerned with some less common methods of integration (by Fubini's theorem, differentiation with respect to a parameter, direct appeal to the definition), with certain methods of setting up integrals which express moments and probabilities, and with the reduction of multiple integrals to single ones. Finally, we consider the role of integration as a source of new functions.

1. INTEGRALS EVALUABLE BY FUBINI'S THEOREM

This method is based on a very special case of Fubini's theorem: if a function $f(x, y)$ is integrated over the rectangle

$$R = \{(x, y): x_0 \leq x \leq x_1, y_0 \leq y \leq y_1\}$$

then under suitable conditions

$$\int_R f(x, y) \, dx \, dy = \int_{x_0}^{x_1} \left(\int_{y_0}^{y_1} f(x, y) \, dy \right) dx = \int_{y_0}^{y_1} \left(\int_{x_0}^{x_1} f(x, y) \, dx \right) dy.$$

Justification is necessary as is shown by the counter-example of $f(x, y) = (x - y)(x + y)^{-3}$:

$$\int_0^1 \int_0^1 f(x, y) \, dy \, dx = 1/2, \qquad \int_0^1 \int_0^1 f(x, y) \, dx \, dy = -1/2.$$

EXAMPLE 1. Let $0 < a \leq b$. Then from

$$\int_0^1 \left(\int_a^b x^y \, dy \right) dx = \int_a^b \left(\int_0^1 x^y \, dx \right) dy$$

we obtain

$$\int_0^1 \frac{x^b - x^a}{\log x}\, dx = \log \frac{b+1}{a+1}.$$

EXAMPLE 2. Let $0 < a \le b$. Then from

$$\int_0^\infty \left(\int_a^b e^{-xy}\, dy \right) dx = \int_a^b \left(\int_0^\infty e^{-xy}\, dx \right) dy$$

we obtain

$$\int_0^\infty (e^{-ax} - e^{-bx}) x^{-1}\, dx = \log \frac{b}{a}.$$

This is a special case of a theorem of Frullani [21]:

$$\int_0^\infty \frac{F(ax) - F(bx)}{x}\, dx = \left[\lim_{u \to \infty} F(u) - F(0) \right] \log \frac{a}{b}$$

if F is suitably integrable and the limit exists. A proof is obtained by starting with

$$\int_0^\infty \left(\int_b^a F'(xy)\, dx \right) dx = \int_b^a \left(\int_0^\infty F'(xy)\, dy \right) dx.$$

EXAMPLE 3. Here we must use an intermediate limit argument. Let $0 < a \le b,\ 0 < L$; then from

$$\int_0^L \left(\int_a^b \sin xy\, dx \right) dy = \int_a^b \left(\int_0^L \sin xy\, dy \right) dx$$

we obtain

$$\int_0^L (\cos by - \cos ay) y^{-1}\, dy = \int_a^b x^{-1} \cos Lx\, dx - \log \frac{b}{a}. \qquad (1)$$

Integration by parts yields

$$\int_a^b x^{-1} \cos Lx\, dx = (Lx)^{-1} \sin Lx \Big|_{x=a}^{x=b} + L^{-1} \int_a^b x^{-2} \sin Lx\, dx$$

so that

$$\lim_{L \to \infty} \int_a^b x^{-1} \cos Lx\, dx = 0.$$

Therefore, passing to the limit on L in (1), we have

$$\int_0^\infty y^{-1} (\cos by - \cos ay)\, dy = \log \frac{a}{b}.$$

It will be seen that a direct argument

$$\int_0^\infty \left(\int_a^b \sin xy \, dx \right) dy = \int_a^b \left(\int_0^\infty \sin xy \, dy \right) dx$$

does not work.

EXAMPLE 4. From

$$\int_0^\infty \left(\int_0^\infty e^{-xy} \sin x \, dy \right) dx = \int_0^\infty \left(\int_0^\infty e^{-xy} \sin x \, dx \right) dy$$

we obtain

$$\int_0^\infty \frac{\sin x}{x} \, dx = \pi/2.$$

A variant of this method occurs when an integral of the type

$$I = \int_a^b f(x)g(x) \, dx$$

is evaluated by substituting another integral for $g(x)$ and then interchanging the order of integrations. For instance, let

$$I = \int_0^\infty \frac{af(x)}{a^2 + x^2} \, dx;$$

since

$$\frac{a}{a^2 + x^2} = \int_0^\infty e^{-au} \cos xu \, du$$

we have

$$I = \int_0^\infty f(x)\left(\int_0^\infty e^{-au} \cos xu \, du \right) dx = \int_0^\infty e^{-au}\left(\int_0^\infty f(x) \cos xu \, dx \right) du.$$

For instance, let $f(x) = \sin x/x$; with some simple trigonometry we show that

$$\int_0^\infty x^{-1} \sin x \cos xu \, dx = \frac{1}{2}\int_0^\infty x^{-1} \sin[x(1 + u)] \, dx - \frac{1}{2}\int_0^\infty x^{-1} \sin[x(1 - u)] \, dx$$

which, by the first integral in the next section, gives us

$$\frac{\pi}{4} [\operatorname{sgn}(1 + u) - \operatorname{sgn}(1 - u)].$$

Therefore

$$\int_0^\infty \frac{a \sin x}{x(a^2 + x^2)} \, dx = \frac{\pi}{4}\int_0^\infty e^{-au}[\operatorname{sgn}(1 + u) - \operatorname{sgn}(1 - u)] \, du = \pi e^{-a}/2a.$$

Using $f(x) = e^{-ax}$, $g(x) = J_0(bx)$, and the Poisson integral for the Bessel function:

$$J_0(bx) = \pi^{-1} \int_0^\pi e^{ibx \cos u} \, du$$

we get in the same way

$$\int_0^\infty e^{-ax} J_0(bx) \, dx = (a^2 + b^2)^{-1/2}.$$

A related method of integration uses the integral transforms. For instance, the Fourier integral formula may be written as

$$\pi f(x) = \int_0^\infty \left(\int_\infty^\infty f(y) \cos[u(y - x)] \, dy \right) du, \qquad \text{general } f,$$

$$\frac{\pi}{2} f(x) = \int_0^\infty \left(\int_0^\infty f(y) \cos uy \, dy \right) \cos ux \, du, \qquad f \text{ even,}$$

$$\frac{\pi}{2} f(x) = \int_0^\infty \left(\int_0^\infty f(y) \sin uy \, dy \right) \sin ux \, du, \qquad f \text{ odd.}$$

Choosing $f(x) = e^{-b|x|}$, with $b > 0$, we have by the middle formula

$$\int_0^\infty \frac{\cos ux}{b^2 + u^2} \, du = \pi e^{-b|x|}/2b;$$

choosing $b > 0$ and

$$f(x) = \begin{matrix} e^{-b|x|} & x > 0 \\ 0 & x = 0 \\ -e^{-b|x|} & x < 0 \end{matrix}$$

we have by the last one of the three formulas

$$\int_0^\infty \frac{u \sin ux}{b^2 + u^2} \, du = \pi e^{-b|x|}/2.$$

Laplace transforms can occasionally be used to evaluate definite integrals. Let

$$F(s) = \int_0^\infty e^{-xs} f(x) \, dx, \qquad G(s) = \int_0^\infty e^{-xs} g(x) \, dx;$$

assuming that the interchange of the orders of integration is allowed, we have

$$\int_0^\infty F(u) g(u) \, du = \int_0^\infty \int_0^\infty e^{-ux} f(x) g(u) \, dx \, du = \int_0^\infty f(u) G(u) \, du.$$

For instance, we have

$$\pi^{1/2}s^{-1/2} = \int_0^\infty e^{-sx}x^{-1/2}\,dx, \qquad e^s E(s) = \int_0^\infty e^{-sx}(1+x)^{-1}\,dx$$

where $E(s)$ is the exponential integral:

$$E(s) = \int_s^\infty e^{-x}x^{-1}\,dx;$$

therefore,

$$\int_0^\infty x^{-1/2}e^x E(x)\,dx = \pi^{1/2}\int_0^\infty x^{-1/2}(1+x)^{-1}\,dx = \pi^{3/2}.$$

More generally, with $0 < b < 1$, we obtain in the same way

$$\int_0^\infty x^{-b}e^x E(x)\,dx = \Gamma(b)\Gamma^2(1-b).$$

It will be seen that all the methods of this section depend essentially on the principle of computing one and the same thing, here a suitable integral, in two different ways.

2. DIFFERENTIATION WITH RESPECT TO A PARAMETER

We have already found $\int_0^\infty x^{-1}\sin x\,dx = \pi/2$. This yields

$$\int_0^\infty \frac{\sin ax}{x}\,dx = \frac{\pi}{2}\,\mathrm{sgn}\,a \tag{1}$$

where the discontinuous factor sgn a is given by

$$\mathrm{sgn}\,a = \begin{matrix} 1 & a > 0 \\ 0 & a = 0 \\ -1 & a < 0. \end{matrix}$$

We prove (1) by letting $f(a)$ be the integral and observing that (a) $f(0) = 0$, (b) $f(-a) = -f(a)$, (c) for $a > 0$, $f(a)$ is a constant independent of a (put $ax = y$), (d) $f(1) = \pi/2$. Suppose now that we wish to evaluate

$$I = \int_0^\infty \frac{\sin^2 x}{x^2}\,dx$$

by differentiation with respect to a parameter. Since there is no parameter we supply one by generalizing our problem to evaluating

$$I(a) = \int_0^\infty \frac{\sin^2 ax}{x^2}\,dx.$$

Differentiating under the integral sign we have

$$I'(a) = \int_0^\infty \frac{\sin 2\,ax}{x}\,dx$$

and since $\operatorname{sgn}(2a) = \operatorname{sgn}(a)$ we have by (1)

$$I'(a) = \frac{\pi}{2}\operatorname{sgn} a$$

whence by integrating

$$I(a) = \frac{\pi}{2}\,|a| + C.$$

Putting $a = 0$ we find that $C = 0$ so that

$$\int_0^\infty \frac{\sin^2 ax}{x^2}\,dx = \frac{\pi}{2}\,|a|,\qquad \int_0^\infty \frac{\sin^2 x}{x^2}\,dx = \frac{\pi}{2}.$$

To evaluate the general case

$$u(m, n) = \int_0^\infty \frac{\sin^m x}{x^n}\,dx \qquad m \geq n$$

we notice the reduction formula

$$(n - 1)(n - 2)u(m, n) + m^2 u(m, n - 2) - m(m - 1)u(m - 2, n - 2) = 0.$$

This together with $u(1, 1) = u(2, 2) = \pi/2$ allows us to compute $u(m, n)$, by some simple trigonometric reduction, for the case when m and n are of the same parity. For instance, we find that

$$u(2j, 2) = \int_0^{\pi/2} \sin^{2j-2} x\,dx$$

and hence we have a formula due to Wolstenholme: if $f(x)$ is analytic and the integrals exist, then [21]

$$\int_0^\infty \frac{\sin^2 x}{x^2} f(\sin^2 x)\,dx = \int_0^{\pi/2} f(\sin^2 x)\,dx.$$

In the case of general m and n we may also proceed as follows. We verify first that

$$u(m, n) = \frac{1}{(n - 1)!} \int_0^\infty \left(\int_0^\infty z^{n-1} e^{-xz} \sin^m x\,dz \right) dx;$$

next, we interchange the order of integrations and use the reduction formula

$$\int_0^\infty e^{-xz} \sin^m x \, dx = \frac{m(m-1)}{m^2 + z^2} \int_0^\infty e^{-xz} \sin^{m-2} x \, dx$$

to obtain

$$u(m, n) = \begin{cases} \dfrac{m!}{(n-1)!} \displaystyle\int_0^\infty \dfrac{z^{n-1} \, dz}{z(z^2 + 2^2)(z^2 + 4^2) \cdots (z^2 + m^2)} & m \text{ even} \\[2em] \dfrac{m!}{(n-1)!} \displaystyle\int_0^\infty \dfrac{z^{n-1} \, dz}{(z^2 + 1^2)(z^2 + 3^2) \cdots (z^2 + m^2)} & m \text{ odd.} \end{cases}$$

Finally, to evaluate the above integrals we use contour integration and residue theorem (or, alternatively, partial fractions).

Suppose next that we wish to evaluate

$$f(a, p) = \int_0^\infty e^{-ax} \frac{\sin px}{x} \, dx. \tag{2}$$

We first reduce this to the smallest possible number of parameters by putting $px = u$; we have then $f(a, p) = g(a/p)$ where

$$g(b) = \int_0^\infty e^{-bx} \frac{\sin x}{x} \, dx.$$

Differentiating with respect to b we find

$$g'(b) = -\int_0^\infty e^{-bx} \sin x \, dx = -1/(1 + b^2)$$

whence by integrating back

$$g(b) = C - \arctan b.$$

With $b = 0$ we find $C = \pi/2$ and so

$$f(a, p) = \frac{\pi}{2} - \arctan \frac{a}{p}.$$

Integrating (2) with respect to a from A to B gives us

$$\int_0^\infty (e^{-Ax} - e^{-Bx}) \frac{\sin px}{x^2} \, dx$$

$$= \frac{\pi}{2}(B - A) - \left(B \arctan \frac{B}{p} - A \arctan \frac{A}{p}\right) + \frac{p}{2} \log \frac{p^2 + B^2}{p^2 + A^2}.$$

To evaluate

$$I_n = \int_0^{\pi/2} (a \cos^2 x + b \sin^2 x)^{-n} \, dx \qquad a, b > 0$$

we have $I_1 = (\pi/2)(ab)^{-1/2}$ and we observe that

$$\left(\frac{\partial}{\partial a} + \frac{\partial}{\partial b} \right) I_n = -n I_{n+1}$$

whence

$$I_{n+1} = \frac{\pi}{2^{2n+1}} (ab)^{-1/2} \sum_{j=0}^{n} \binom{2j}{j} \binom{2n-2j}{n-j} a^{-j} b^{j-n}.$$

To evaluate

$$F(a) = \int_0^a \frac{\log(1 + ax)}{1 + x^2} \, dx$$

we differentiate with respect to a and get

$$F'(a) = \frac{\log(1 + a^2)}{1 + a^2} + \int_0^a \frac{x \, dx}{(1 + ax)(1 + x^2)}.$$

This is evaluated by partial fractions:

$$F'(a) = \frac{\log(1 + a^2)}{2(1 + a^2)} + \frac{a}{1 + a^2} \arctan a,$$

now we integrate:

$$F(a) = \int_0^a \frac{x \arctan x}{1 + x^2} \, dx + \frac{1}{2} \int_0^a \frac{\log(1 + x^2)}{1 + x^2} \, dx. \tag{3}$$

Integrating by parts we find

$$\int_0^a \frac{\log(1 + x^2)}{1 + x^2} \, dx = (\arctan a)\log(1 + a^2) - 2 \int_0^a \frac{x \arctan x}{1 + x^2} \, dx$$

and substituting this into (3) we have

$$F(a) = \frac{1}{2} (\arctan a)\log(1 + a^2).$$

In particular, putting $a = 1$ we have

$$\int_0^1 \frac{\log(1 + x)}{1 + x^2} \, dx = \frac{\log 2}{8} \pi.$$

Sometimes we evaluate a definite integral by showing that it satisfies a differential equation. Consider for example

$$F(a, b) = \int_0^\infty e^{-ax^2} \cos bx \, dx;$$

we reduce this to one parameter by putting $x = a^{-1/2}u$ and we have

$$F(a, b) = a^{-1/2}f(t), \qquad f(t) = \int_0^\infty e^{-x^2} \cos tx, \, dx, \qquad t = ba^{-1/2}.$$

To evaluate $f(t)$ we differentiate with respect to t and then integrate by parts to obtain

$$f'(t) = -\frac{t}{2}f(t).$$

Solving the differential equation we have

$$f(t) = Ce^{-t^2/4}$$

and $C = \pi^{1/2}/2$ since $f(0) = \pi^{1/2}/2$. Therefore

$$\int_0^\infty e^{-ax^2} \cos bx \, dx = (\pi/4a)^{1/2} e^{-b^2/4a}.$$

Replacing b by $2b$ and observing that

$$\cos 2bx = \cos^2 bx - \sin^2 bx, \qquad \int_0^\infty e^{-ax^2} \, dx = (\pi/4a)^{1/2}$$

we derive the integrals

$$\int_0^\infty e^{-ax^2} \cos^2 bx \, dx = (\pi/4a)^{1/2} \frac{1 + e^{-b^2/a}}{2},$$

$$\int_0^\infty e^{-ax^2} \sin^2 bx \, dx = (\pi/4a)^{1/2} \frac{1 - e^{-b^2/a}}{2}.$$

We note several other ways of evaluating $F(a, b)$: (a) by complex variable methods (based on $\cos bx = \operatorname{Re} e^{ibx}$), (b) expanding $\cos bx$ in powers of x and integrating term-by-term, (c) by starting with

$$G(a, b) = \int_0^\infty e^{-ax^2} \cosh bx \, dx = \frac{1}{2}\int_{-\infty}^\infty e^{-ax^2} \cosh bx \, dx;$$

we have then by completing squares

$$G(a, b) = \frac{1}{4} e^{b^2/4a} \left[\int_{-\infty}^{\infty} e^{-a(x-b/2a)^2} \, dx + \int_{-\infty}^{\infty} e^{-a(x+b/2a)^2} \, dx \right]$$

$$= \frac{1}{2} e^{b^2/4a} \int_{-\infty}^{\infty} e^{-ax^2} \, dx = (\pi/4a)^{1/2} e^{b^2/4a}$$

and we observe that

$$F(a, b) = G(a, ib) = (\pi/4a)^{1/2} e^{-b^2/4a}.$$

Finally, replacing in $F(a, b)$ the real parameter a by the complex parameter $\alpha + i\beta$, and separating the real and imaginary parts we obtain for $\alpha \geq 0$ (paying some attention to signs of roots of complex numbers)

$$\int_0^{\infty} e^{-\alpha x^2} \cos \beta x^2 \cos bx \, dx$$

$$= \frac{1}{2} \pi^{1/2} (\alpha^2 + \beta^2)^{-1/4} e^{-b^2\alpha/[4(\alpha^2 + \beta^2)]} \cos \left[\frac{b^2\beta}{4(\alpha^2 + \beta^2)} - \frac{1}{2} \arctan \frac{\beta}{\alpha} \right]$$

$$\int_0^{\infty} e^{-\alpha x^2} \sin \beta x^2 \cos bx \, dx$$

$$= \frac{1}{2} \pi^{1/2} (\alpha^2 + \beta^2)^{-1/4} e^{-b^2\alpha/[4(\alpha^2 + \beta^2)]} \sin \left[\frac{1}{2} \arctan \frac{\beta}{\alpha} - \frac{b^2\beta}{4(\alpha^2 + \beta^2)} \right].$$

In particular, with $\alpha = b = 0$ we get the Fresnel integrals

$$\int_0^{\infty} \sin \beta x^2 \, dx = \int_0^{\infty} \cos \beta x^2 \, dx = \frac{1}{2} \sqrt{\frac{\pi}{2\beta}}.$$

Differentiation with respect to a parameter together with suitable use of complex numbers will sometimes yield us complicated integrals in a straightforward way. For instance, by the definition of Γ-function we have

$$\int_0^{\infty} t^{a-1} e^{-pt} \, dt = \Gamma(a) p^{-a}; \qquad (4)$$

differentiating k times with respect to p we find that

$$\int_0^{\infty} t^{a-1} t^k e^{-pt} \, dt = \Gamma(a) a(a + 1) \cdots (a + k - 1) p^{-a-k}.$$

Multiplying both sides by $(iq)^k/k!$ and summing on k from 0 to ∞ gives us by the binomial theorem

$$\int_0^{\infty} t^{a-1} e^{iqt} e^{-pt} \, dt = \Gamma(a)(p - iq)^{-a}$$

which we could also get directly from (4) by replacing p with $p - iq$. Continuing, we have

$$\int_0^\infty t^{a-1} e^{iqt} e^{-pt}\, dt = \Gamma(a)(p^2 + q^2)^{-a/2} e^{ia \text{ arc tan } q/p};$$

separating the real and imaginary parts we get

$$\int_0^\infty t^{a-1} \cos qt\; e^{-pt}\, dt = \Gamma(a)(p^2 + q^2)^{-a/2} \cos (a \text{ arc tan } q/p);$$

$$\int_0^\infty t^{a-1} \sin qt\; e^{-pt}\, dt = \Gamma(a)(p^2 + q^2)^{-a/2} \sin (a \text{ arc tan } q/p).$$
(5)

Differentiating k times with respect to a, we find the integrals

$$\int_0^\infty t^{a-1}(\log t)^k \cos qt\; e^{-pt}\, dt, \qquad \int_0^\infty t^{a-1}(\log t)^k \sin qt\; e^{-pt}\, dt.$$

Putting $p = 0$ in (5) we have

$$\int_0^\infty t^{a-1} \cos qt\, dt = q^{-a}\Gamma(a) \cos \frac{\pi a}{2},$$

$$\int_0^\infty t^{a-1} \sin qt\, dt = q^{-a}\Gamma(a) \sin \frac{\pi a}{2}.$$

Finally, replacing a by $a + 1$ in (5), and then writing $a + ib$ in place of a and separating the real and imaginary parts, we find integrals of the type

$$\int_0^\infty t^a \cos (b \log t) \cos qt\; e^{-pt}\, dt.$$

3. INTEGRALS EVALUABLE FROM FIRST PRINCIPLES

It is sometimes possible to evaluate an integral by an appeal to its definition as the limit of a sum. Consider for instance

$$I = \int_0^\pi \log \sin x\, dx.$$

It is proved in the chapter on series and products that

$$\prod_{j=1}^{n-1} \sin \frac{j\pi}{n} = n \cdot 2^{1-n}$$

whence

$$\sum_{j=1}^{n-1} \log \sin \frac{j\pi}{n} = \log n - (n - 1)\log 2.$$
(1)

By its definition as the limit

$$I = \lim_{n \to \infty} \left[\frac{\pi}{n} \sum_{j=1}^{n-1} \log \sin \frac{j\pi}{n} \right]$$

so that by (1) $I = -\pi \log 2$. Another way to evaluate I is the following:

$$\frac{I}{2} = \int_0^{\pi/2} \log \sin x \, dx = \int_0^{\pi/2} \log \cos x \, dx$$

so that

$$I = \int_0^{\pi/2} (\log \sin x + \log \cos x) \, dx = \int_0^{\pi/2} \log \frac{\sin 2x}{2} \, dx$$

$$= \int_0^{\pi/2} \log \sin 2x \, dx - \frac{\pi}{2} \log 2.$$

The substitution $2x = y$ shows that

$$\int_0^{\pi/2} \log \sin 2x \, dx = I/2$$

so that

$$I = \frac{I}{2} - \frac{\pi}{2} \log 2$$

whence again $I = -\pi \log 2$. Integrating by parts we find that

$$\int_0^{\pi/2} x \cot x \, dx = \frac{\pi}{2} \log 2,$$

substituting $\sin x = u$ and then $u = e^{-y}$ we find

$$\int_0^1 \frac{\log u}{\sqrt{1 - u^2}} \, du = -\frac{\pi}{2} \log 2, \qquad \int_0^\infty \frac{y \, dy}{\sqrt{e^{2y} - 1}} = \frac{\pi}{2} \log 2.$$

Similarly, starting with the factorization

$$r^{2m} - 2r^m \cos m\theta + 1 = \prod_{j=0}^{m-1} \left[r^2 - 2r \cos\left(\theta + \frac{2j\pi}{m}\right) + 1 \right],$$

putting $\theta = 0$, and taking logarithms, we find that

$$J = \int_0^{2\pi} \log(r^2 - 2r \cos \theta + 1) \, dr = \lim_{m \to \infty} \frac{2\pi}{m} \log(r^m - 1)^2$$

so that

$$J = J(r) = \begin{cases} 0, & 0 \le r < 1 \\ 4\pi \log r, & 1 < r. \end{cases}$$

This integral could also be evaluated by a special device. First, we show that

$$J(r) + J(-r) = J(r^2), \qquad J(r) = J(-r);$$

it follows that

$$J(r) = 2^{-n} J(r^{2^n}).$$

Hence, passing to the limit on n, we find that

$$J(r) = 0, \qquad 0 \le r \le 1;$$

notice that the case of $J(1)$ was not settled by the previous argument. If $r > 1$ put $r = 1/s$ so that $s < 1$, and we have

$$J(s) = J(r) - 4\pi \log r.$$

Since $J(s) = 0$ it follows that $J(r) = 4\pi \log r$.

4. DILOGARITHMS

It is simple to derive certain properties of the logarithm from its definition as an integral:

$$\log x = \int_1^x \frac{du}{u}.$$

For instance, breaking up the interval of integration and using a substitution, we have

$$\log(xy) = \int_1^{xy} \frac{du}{u} = \int_1^x \frac{du}{u} + \int_x^{xy} \frac{du}{u} = \int_1^x \frac{du}{u} + \int_1^y \frac{dv}{v} = \log x + \log y.$$

Here we shall consider a non-elementary function related to the logarithm. The dilogarithm may be defined by the series

$$L(z) = \frac{z}{1^2} + \frac{z^2}{2^2} + \frac{z^3}{3^2} + \cdots \tag{1}$$

which converges for $|z| \le 1$. Proceeding as in the section on the summation of series, we express it as an integral (we sum it in closed form):

$$L(z) = -\int_0^z \frac{\log(1-x)}{x}\, dx. \tag{2}$$

There also exists the general polylogarithm $L_k(z)$:

$$L_k(z) = \sum_{n=1}^{\infty} z^n/n^k,$$

with similar integral representation:

$$L_k(z) = \int_0^z \frac{1}{z_1} \int_0^{z_1} \frac{1}{z_2} \cdots \int_0^{z_{k-2}} \frac{1}{z_{k-1}} [-\log(1 - z_{k-1})]\, dz_{k-1} \cdots dz_2\, dz_1.$$

While the series (1) is defined only for $|z| \leq 1$ the integral (2), with some care as to the branch of the logarithm, is defined for every complex z. By differentiating (2) we find

$$\frac{d}{dx} L(-1/z) = [\log(1 + x) - \log x]/x$$

and by integrating it back

$$L(-1/x) + L(-x) + \frac{1}{2} \log^2 x = C = 2L(-1) \tag{3}$$

the value of C being obtained by putting $x = 1$. Since by (1)

$$-2L(-1) = L(1)$$

putting $x = e^{\pi i}$ in (3) we obtain

$$L(1) = \sum_{n=1}^{\infty} 1/n^2 = \pi^2/6, \qquad L(-1) = -\pi^2/12.$$

Starting with (2) and integrating by parts we have (exercising some care)

$$L(z) = -\log z \log(1 - z) - L(1 - z) + L(1)$$

or

$$L(z) + L(1 - z) = \pi^2/6 - \log z \log(1 - z). \tag{4}$$

Similarly to (3), we have

$$\frac{d}{dx} L\left(-\frac{x}{1 - x}\right) = \left(\frac{1}{x} + \frac{1}{1 - x}\right)\log(1 - x)$$

and by integrating

$$L(x) + L\left(-\frac{x}{1 - x}\right) = -\frac{1}{2} \log^2(1 - x), \qquad x < 1. \tag{5}$$

4. DILOGARITHMS

Starting with the factorization

$$1 - x^r = \prod_{j=0}^{r-1}(1 - xe^{2\pi ij/r}),$$

taking logarithms, dividing by x, and integrating, we have

$$\frac{1}{r}L(x^r) = \sum_{j=0}^{r-1}L(xe^{2\pi ij/r});$$

in particular for $r = 2$

$$\frac{1}{2}L(x^2) = L(x) + L(-x). \tag{6}$$

As already found, $L(1) = \pi^2/6$ and $L(-1) = -\pi^2/12$; putting $z = 1/2$ in (4) we get

$$L(1/2) = \pi^2/12 - \frac{1}{2}\log^2 2.$$

Eliminating $L(x)$ out of (5) and (6) gives us

$$L\left(-\frac{x}{1-x}\right) + \frac{1}{2}L(x^2) - L(-x) = -\frac{1}{2}\log^2(1-x); \tag{7}$$

we equate the arguments by putting

$$-\frac{x}{1-x} = x^2 \quad \text{or} \quad x^2 - x - 1 = 0$$

so that

$$x = \frac{1 - \sqrt{5}}{2}$$

since (to apply (5)) we need the root < 1. Now we have (7) as

$$\frac{3}{2}L\left(\frac{3-\sqrt{5}}{2}\right) - L\left(\frac{\sqrt{5}-1}{2}\right) = -\frac{1}{2}\log^2\left(\frac{\sqrt{5}+1}{2}\right);$$

further, putting $z = (3 - \sqrt{5})/2$ in (4) we obtain

$$L\left(\frac{3-\sqrt{5}}{2}\right) + L\left(\frac{\sqrt{5}-1}{2}\right) = \pi^2/6 - \log\frac{3-\sqrt{5}}{2}\log\frac{\sqrt{5}-1}{2}.$$

The last two equations can be solved for $L[(3 - \sqrt{5})/2]$ and $[(\sqrt{5} - 1)/2]$ and we get

$$L\left(\frac{3 - \sqrt{5}}{2}\right) = \pi^2/15 - \frac{1}{4}\log^2\frac{3 - \sqrt{5}}{2}$$

$$L\left(\frac{\sqrt{5} - 1}{2}\right) = \pi^2/10 - \log^2\frac{\sqrt{5} - 1}{2}.$$

Putting $x = (3 - \sqrt{5})/2$ and $x = (\sqrt{5} - 1)/2$ in (5) gives us

$$L\left(\frac{1 - \sqrt{5}}{2}\right) = -\pi^2/15 + \frac{1}{2}\log^2\frac{\sqrt{5} - 1}{2},$$

$$L\left(-\frac{1 + \sqrt{5}}{2}\right) = -\pi^2/10 + \frac{1}{2}\log^2\frac{\sqrt{5} + 1}{2}.$$

We notice here yet another appearance of the ubiquitous "golden ratio" number $(1 + \sqrt{5})/2$. It appears to be unknown whether the dilogarithm $L(x)$ can be explicitly expressed in a finite form for any further values of x. For a general reference to dilogarithms and related functions see [42].

5. CROFTON'S THEOREM AND CONFIGURATION AVERAGES

Suppose that C is a configuration consisting of n points taken independently and uniformly at random in a region R of some Euclidean space E^m, and let the volume of R be V. Let $F(C)$ be a real-valued function associated with C, such as some length, area or volume, or a discontinuous factor which is 1 if C satisfies a certain configuration condition and which is 0 otherwise. The sole requirement is that $F(C)$ is invariant under rigid motions of C as a whole, so that it does not depend on R or on the orientation of C with respect to R. We are interested in computing the average M:

$$M = V^{-n}\int_R \cdots \int_R F(C)\,dx_1 \cdots dx_n. \tag{1}$$

Crofton's theorem relates M to another quantity M_1 which is the average of $F(C)$ when $n - 1$ points vary independently over R while one point varies over the boundary ∂R of R:

$$M_1 = V^{-(n-1)}[\text{Content }(\partial R)]^{-1}\int_R \cdots \int_R \int_{\partial R} F(C)\,dx_1 \cdots dx_{n-1}\,dx_n.$$

A sketch of the derivation of Crofton's theorem follows. Let $R + \Delta R$ be a larger region, containing R, and of volume $V + \Delta V$. Denoting by I the integral in (1) we have

$$I = V^n M, \qquad I + \Delta I = (V + \Delta V)^n(M + \Delta M). \tag{2}$$

For the augmented integral $I + \Delta I$ we have

$$I + \Delta I = I_0 + I_1 + I_2 + \cdots + I_n$$

where $I_0 = I$ and I_j is the contribution arising from the case when $n - j$ points are in R and j in ΔR. Under some minimal assumptions of sufficient regularity we have $I_j = 0(\Delta V)^j$ and in particular

$$I_1 = nM_1 V^{n-1}\Delta V + 0(\Delta V)^2; \tag{3}$$

here the factor n occurs because any one of the n points may be the one in ΔR. Now from (2) and (3) we obtain

$$\Delta M = nV^{-1}(M_1 - M)\Delta V + 0(\Delta V)^2 \tag{4}$$

which expresses Crofton's theorem. In practice (4) is used to derive a differential equation for M, once M_1 is known.

EXAMPLE 1. Two points are taken independently and uniformly at random inside or on a circle of radius r; k is a real number with $k > -2$; what is the average $M_k(r)$ of the kth power of the distance between the points? This could be set up directly as a quadruple integral

$$M_k(r) = \pi^{-2}r^{-4} \int_0^{2\pi}\int_0^{2\pi}\int_0^r\int_0^r [r_1{}^2 + r_2{}^2 - 2r_1 r_2 \cos(\theta_1 - \theta_2)]^{k/2}$$
$$\times r_1 r_2\, dr_1\, dr_2\, d\theta_1\, d\theta_2$$

but it is easier to proceed via Crofton's theorem. We first compute the corresponding average $m_k(r)$ when one of the two points is on the boundary. By symmetry, we can fix the boundary point. With reference to Figure 1 we find that the elementary area shown shaded is $2\theta x\, dx$ where $\cos\theta = x/2r$, whence

$$m_k(r) = \frac{2}{\pi r^2} \int_0^{2r} x^{k+1} \arccos\frac{x}{2r}\, dx$$

so that letting $\theta = \arccos x/2r$ and integrating by parts

$$m_k(r) = \frac{2^{2k+4}\Gamma^2\left(\dfrac{k+3}{2}\right)}{(k+2)\Gamma(k+3)} r^k.$$

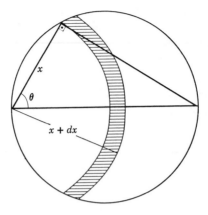

FIGURE 1.   Area element for $m_k(r)$.

Next, we apply Crofton's theorem using as the enlarged region a concentric circle of radius $r + \Delta r$. Then (4) yields

$$\Delta M_k(r) = \frac{4}{r} [m_k(r) - M_k(r)]\Delta r + 0(\Delta r)^2$$

and by passage to the limit we have the differential equation

$$M_k'(r) = 4r^{-1}m_k(r) - 4r^{-1}M_k(r).$$

Integrating and observing that $M_k(0) = 0$ we have

$$M_k(r) = \frac{2^{2k+6}\Gamma^2\left(\dfrac{k+3}{2}\right)}{(k+4)(k+2)\Gamma(k+3)\pi} r^k.$$

In particular, $M_{-1}(r) = (16/3\pi)r^{-1}$, $M_1(r) = (128/45\pi)r$, $M_2(r) = r^2$, $M_3(r) = (2048/525\pi)r^3$.

More generally, given a plane closed convex curve K, we consider a random chord, and the distance between a pair of random points inside K. Let $I_k(K)$ and $J_k(K)$ be the integrals of the kth powers of the chord length and of the distance; then

$$J_k(K)/\text{Area}^2(K)$$

generalizes the quantity $M_k(r)$ from circles of radius r to arbitrary convex curves. By combining the Crofton method with those of the section on integral geometry in the chapter on geometry, one can prove that

$$J_k(K) = \frac{2}{(k+2)(k+3)} I_{k+3}(K).$$

and since $J_0(K) = \text{Area}^2(K)$ we obtain another formula of Crofton

$$I_3(K) = 3 \text{ Area}^2(K).$$

Slightly differently put, this asserts that the mean of the cube of the length of a random chord of K is three times the square of the area enclosed by K, divided by the length of K.

We have used Crofton's method to find the average $M_k(r)$ of the kth power of the distance between two points inside the circle of radius r. It is also possible to find in the same way the probability density $p(x, r)$; $p(x, r) \, dx$ is now the probability that the distance in question lies between x and $x + dx$. For if $p_1(x, r)$ is the same density when one of the points is on the circle then by reference to Figure 1 we have

$$p_1(x, r) = 2\theta x/\pi r^2 \tag{5}$$

and by Crofton's formula (4)

$$dp = 4(p_1 - p) \, dr/r. \tag{6}$$

Here, in augmenting the domain from a circle of radius r to one of radius $r + dr$, we have to keep x constant, hence $x = 2r \cos \theta$ leads to the relation

$$\tan \theta \, d\theta = dr/r.$$

Therefore (6) leads to the differential equation

$$\frac{dp}{d\theta} + 4p \tan \theta = \frac{32\theta}{\pi x} \cos \theta \sin \theta$$

whose solution is

$$p = \frac{16 \cos^2 \theta}{\pi x} (\theta - \sin \theta \cos \theta), \qquad \theta = \text{arc} \cos \frac{x}{2r}.$$

EXAMPLE 2.   We consider a special case of Sylvester's four points-problem: given four points uniformly and at random in a square, what is the probability p that the quadrilateral Q determined by them is convex? This is handled by Crofton's theorem and here the configuration function F is 1 or 0 depending on whether Q is convex or not. If Q is concave then one of the four points lies in the triangle formed by the other three, hence

$$p = 1 - 4T/S$$

where S is the area of the square and T is the mean area of a triangle formed by three random points in the square. By Crofton's theorem we reduce finding T to the case when one of the three points is in the boundary, i.e., is taken at random along a side of the square. Observing that any one of the three points may be the boundary one, the reader may wish to show that $T = 11S/72$ and hence $p = 25/36$. Further, since the convexity of Q remains unchanged under projections, $p = 25/36$ for four points in a rectangle and, more generally, in a parallelogram. For a triangle the corresponding value is $p = 2/3$ and for a circle $p = 1 - 35/12\pi^2 = 0.70405$. Thus the probability $p(R)$, considered as a function of the convex region R in which the four points lie, is a measure of symmetry of R. It appears to be still unknown whether for any convex R,

$$p(R) \leq 1 - 35/12\pi^2,$$

the corresponding lower bound, 2/3, is known.

6. FORBIDDEN CONFIGURATIONS AND DISCONTINUOUS FACTORS

Certain probabilities of the combinatorial-geometrical type are expressible in the form

$$P = \int_A f(x)\, dx \tag{1}$$

where the vector variable x ranges over a complicated subset of the Euclidean space E^n. Let $\chi_A(x)$ be the characteristic function of A, then (1) may be written

$$P = \int_{E^n} f(x)\chi_A(x)\, dx \tag{2}$$

which appears merely to have transferred the difficulty from the domain of integration to the integrand. However, an integral representation

$$\chi_A(x) = \int_G F(x, u)\, du$$

can be sometimes be found; then, substituting into (2) and interchanging the integrations, we have

$$P = \int_G \int_{E^n} f(x)F(x, u)\, dx\, du$$

which may be more tractable than (1).

EXAMPLE 1.   Four labeled points $p_i = (x_i, y_i)$, $i = 1, 2, 3, 4$, are taken in a plane convex region R independently, according to some probability law. Express analytically the probabilities P_1, P_2, P_3 that the closed quadrilateral $p_1 p_2 p_3 p_4 p_1$ is (a) convex, (b) concave, (c) self-intersecting.

The equation of the straight line L_{ij} through p_i and p_j is

$$(x_j - x_i)y - (y_j - y_i)x + x_i y_j - y_i x_j = 0$$

which we write as $L_{ij}(x, y) = 0$. It follows that

$$L_{13}(x_2, y_2)L_{13}(x_4, y_4) \tag{3}$$

is positive if p_2 and p_4 lie on the same side of L_{13}, and negative if they are separated by L_{13}; we omit the 0-probability event that some three points are collinear. For brevity we write (3) as

$$L_{13}{}^2 L_{13}{}^4.$$

Now we verify that the quadrilateral is

$$
\begin{array}{ll}
\text{convex if} & L_{13}{}^2 L_{13}{}^4 < 0 \quad \text{and} \quad L_{24}{}^1 L_{24}{}^3 < 0 \\
\text{concave if} & \text{one of these is} < 0 \text{ and one} > 0 \\
\text{self-crossing if} & L_{13}{}^2 L_{13}{}^4 > 0 \quad \text{and} \quad L_{24}{}^1 L_{24}{}^3 > 0.
\end{array}
$$

Using this and the discontinuous factor sgn x we can write down the three characteristic functions

$$\chi_1 = \frac{1}{4}[1 - \text{sgn}(L_{13}{}^2 L_{13}{}^4)][1 - \text{sgn}(L_{24}{}^1 L_{24}{}^3)],$$

$$\chi_2 = \frac{1}{2}[1 - \text{sgn}(L_{13}{}^2 L_{13}{}^4 L_{24}{}^1 L_{24}{}^3)],$$

$$\chi_3 = \frac{1}{4}[1 + \text{sgn}(L_{13}{}^2 L_{13}{}^4)][1 + \text{sgn}(L_{24}{}^1 L_{24}{}^3)]$$

where $\chi_1 = 1$ if our quadrilateral is convex and $\chi_1 = 0$ otherwise, while χ_2 and χ_3 perform similarly for concave and self-crossing cases. Now the probabilities P_i are the integrals of the χ_i with respect to the probability density governing the choice of points in R.

If the four points are taken uniformly at random in R then the evaluation of P_i's is easiest by Sylvester's problem of the previous section. We have then $P_1 + P_3 = $ Prob (the quadrilateral determined by the four *unlabeled* points is convex) $= p$, say, which is evaluable as before by Crofton's theorem, and $P_3 = 2P_1$ because out of 24 equally likely permutations of the four points 8 lead to the convex case and 16 to the self-crossed case. Hence $P_1 = p/3$, $P_2 = 1 - p$, $P_3 = 2p/3$.

EXAMPLE 2. Six labeled points $p_i = (x_i, y_i, z_i)$, $i = 1, \ldots, 6$, are taken independently at random in some region R of E^3, according to a prescribed probability law. Express analytically the probability P that the closed piecewise linear path $p_1 p_2 p_3 p_4 p_5 p_6 p_1$ forms a knot.

It will be convenient to have open-path knots as well as the usual closed ones. The open rectilinear path $p_1p_2p_3p_4p_5p_6$ is an open knot if on extending the last and first segments indefinitely in a straight line (p_1p_2 beyond p_1, p_5p_6 beyond p_6) we obtain a knot tied on the straight line. We observe that the only closed knot $p_1 \ldots p_6p_1$ possible is the usual trefoil knot. By considering over- and under-passes it is easy to see that no 5-point rectilinear path $p_1p_2p_3p_4$ p_5 can form an open knot. Hence no closed 5-point path can form a closed knot, for otherwise the removal of one of the five edges would have resulted in an open knot. Further, if the closed path $p_1 \ldots p_6p_1$ forms a knot then one edge can be separated from the other four points by a plane, and the remaining five edges form an open knot. Suppose for the time being that the single edge so separated is p_1p_6.

Given three consecutive vertices of the path, say, p_1, p_2, p_3, let $[p_1p_2p_3]$ denote the smaller of the two angular regions into which the plane through p_1, p_2, p_3 is divided by the straight rays from p_2 which pass through p_1 and p_3. Recalling the separation condition on p_1p_6, we find that the path $p_1 \ldots p_6$ forms an open knot if and only if:

$$p_1p_2 \text{ cuts } [p_4p_5p_6] \text{ and } p_5p_6 \text{ cuts } [p_1p_2p_3] \text{ and } p_5p_6 \text{ cuts } [p_2p_3p_4].$$

The condition that p_1p_2 cuts $[p_4p_5p_6]$ may be written more conveniently: the plane through p_1, p_2 and p_5 separates p_4 from p_6. The equation of that plane is

$$\begin{vmatrix} 1 & 1 & 1 & 1 \\ x_1 & x_2 & x_5 & x \\ y_1 & y_2 & y_5 & y \\ z_1 & z_2 & z_5 & z \end{vmatrix} = 0$$

which we write briefly as $D_{125}(x, y, z) = 0$; denote by D_{125}^4 the result of replacing x, y, z in $D_{125}(x, y, z)$ by x_4, y_4, z_4. Now the necessary and sufficient condition for an open knot is:

$$D_{125}^4 D_{125}^6 < 0 \quad \text{and} \quad D_{256}^1 D_{256}^3 < 0 \quad \text{and} \quad D_{356}^2 D_{356}^4 < 0.$$

The corresponding characteristic function is

$$\chi_1 = \chi_1(x_1, y_1, \ldots, z_6)$$

$$= \frac{1}{8}[1 - \text{sgn}(D_{125}^4 D_{125}^6)][1 - \text{sgn}(D_{256}^1 D_{256}^3)][1 - \text{sgn}(D_{356}^2 D_{356}^4)]; \quad (4)$$

this is 1 for a knot and 0 otherwise. We return now to the separation condition: it was assumed that p_1p_6 is separated from the other four points; what if the whole configuration is a knot but p_1p_6 is not the separable edge? By considering planes sufficiently close to p_ip_{i+1} which are parallel to planes through some three points p_j, with $j \neq i, j \neq i + 1$, we can show that the configuration is a knot if and only if three edges can be separated. Further, if the configuration is not a closed knot, then at most one edge can be removed, with the five others forming an open knot.

Suppose now that $\chi_2, \ldots, \chi_6$ are obtained from χ_1 of (4) by cyclic permutations of the six points $p_1, \ldots, p_6$. It follows that

$$\chi = \chi(x_1, y_1, \ldots, z_6) = \frac{1}{3} \sum_{1 \le i < j \le 6} \chi_i \chi_j$$

is the characteristic function of the knot: 1 if the configuration is a knot, 0 otherwise. The probability P is easily expressible as an integral involving χ.

7. MISCELLANEOUS

Consider the integral

$$I(a) = \int_0^\infty \frac{\log x}{(a + x)^2} \, dx, \qquad a > 0;$$

this cannot be evaluated by a straightforward integration by parts, and we proceed as follows. Put $x = ay$, then

$$I(a) = \frac{\log a}{a} \int_0^\infty \frac{dy}{(1 + y)^2} + \frac{1}{a} I(1);$$

the integral here is equal to 1 and, by splitting the range of integration in $I(1)$ into the intervals $[0, 1]$ and $[1, \infty)$ and using the substitution $x = 1/y$ on $[0, 1]$, we show that $I(1) = 0$. Hence

$$\int_0^\infty \frac{\log x}{(a + x)^2} \, dx = \frac{\log a}{a}. \tag{1}$$

Differentiating n times with respect to a we have

$$\int_0^\infty \frac{\log x}{(a + x)^{n+2}} \, dx = \frac{1}{n + 1} a^{-n-1} \left[\log a - \sum_{j=1}^n 1/j \right].$$

Putting $a = n$ and rearranging we find that

$$\int_0^\infty \frac{(n + 1)n^{n+1}}{(n + x)^{n+2}} \log x \, dx = \log n - \sum_{j=1}^n 1/j$$

and passing to the limit on n we get

$$\int_0^\infty e^{-x} \log x \, dx = -\gamma,$$

the Euler constant. Similarly, starting with the integral

$$J(a) = \int_0^\infty \frac{\log^2 x}{(a + x)^2} \, dx$$

and proceeding as before, we find

$$J(a) = \frac{\log^2 a}{a} + \frac{1}{a} J(1).$$

Here $J(1) = \pi^2/3$ so that

$$\int_0^\infty \frac{\log^2 x}{(a + x)^2} \, dx = \frac{\log^2 a}{a} + \pi^2/3a. \tag{2}$$

Differentiating n times with respect to a, we find

$$\int_0^\infty \frac{\log^2 x}{(a + x)^{n+2}} \, dx = \frac{1}{n+1} a^{-n-1} \left[\frac{\pi^2}{3} - \sum_{j=1}^n 1/j^2 + \left(\log a - \sum_{j=1}^n 1/j \right)^2 \right].$$

Putting as before $a = n$, rearranging, and passing to the limit on n, we have

$$\int_0^\infty e^{-x} \log^2 x \, dx = \frac{\pi^2}{6} - \gamma^2.$$

If instead of differentiating in (1) and (2) with respect to a, we integrate with respect to a, from 1 to u, we find

$$\int_0^\infty \left(\frac{1}{1 + x} - \frac{1}{u + x} \right) \log x \, dx = \frac{1}{2} \log^2 u,$$

$$\int_0^\infty \left(\frac{1}{1 + x} - \frac{1}{u + x} \right) \log^2 x \, dx = \frac{1}{3} \log^3 u + \frac{\pi^2}{3} \log u.$$

Besides differentiating and integrating with respect to a parameter it is occasionally useful to apply the operation of differencing:

if $f(u) = \int F(x, u) \, dx$ then $f(u + 1) - f(u) = \int [F(x, u + 1) - F(x, u)] \, dx$.

For instance, starting with

$$\int_0^\infty e^{-xu} \, dx = 1/u$$

and differencing n times we find

$$\int_0^\infty e^{-ux}(1 - e^{-x})^n \, dx = \frac{1}{u(u + 1) \cdots (u + n)}. \tag{3}$$

A series of the form

$$F(u) = \sum_{n=0}^{\infty} \frac{a_n}{u(u+1)\cdots(u+n)}$$

is called a factorial series; if

$$f(x) = \sum_{n=0}^{\infty} a_n x^n$$

is the corresponding power series then by (3) we have

$$F(u) = \int_0^{\infty} e^{-ux} f(1 - e^{-x})\, dx$$

which can be written, on putting $1 - e^{-x} = t$, as

$$F(u) = \int_0^1 (1-t)^{u-1} f(t)\, dt.$$

The integrals

$$\int_0^1 x^{a-cx}\, dx, \qquad \int_0^1 x^{-cx^a}\, dx$$

are easily evaluated by writing the integrands as exponentials

$$e^{(a-cx)\log x}, \qquad e^{-cx^a \log x},$$

substituting $x = e^{-u}$, expanding by the exponential series, and integrating term-by-term; we find thus

$$\int_0^1 x^{a-cx}\, dx = \sum_{n=0}^{\infty} \frac{c^n}{(a+n+1)^{n+1}}, \tag{4}$$

$$\int_0^1 x^{-cx^a}\, dx = \sum_{n=0}^{\infty} \frac{c^n}{(1+na)^{n+1}}. \tag{5}$$

In particular, putting $a = 0$ and $c = 1$ in (4) or $a = c = 1$ in (5) we have

$$\int_0^1 \frac{dx}{x^x} = \sum_{n=1}^{\infty} \frac{1}{n^n},$$

and putting $a = 0$ and $c = -1$ in (4) or $a = 1$ and $c = -1$ in (5) we get

$$\int_0^1 x^x\, dx = \sum_{n=1}^{\infty} \frac{(-1)^{n+1}}{n^n}.$$

Putting $a = (p-1)/q$ and $c = 1/q$ in (5) and summing over q gives us

$$\int_0^1 \left(\sum_{q=1}^{p-1} q^{-1} x^{-\frac{1}{q} x^{(p-1)/q}} \right) dx = \sum_{n=1}^{\infty} \frac{1}{n^{1+[n/p]}}$$

where, as usual, $[x]$ denotes the greatest integer $\leq x$. The following transformation of infinite integrals is due to Schlömilch. Let

$$I = \int_0^\infty f[(cx - a/x)^2]\, dx, \qquad a, c > 0.$$

be a well-defined convergent integral. Put

$$F(x) = \frac{1}{2c}\,(c + a/x^2)[1 + (cx - a/x)(4ac + (cx - a/x)^2)^{-1/2}],$$

then it is easily verified that $F(x) \equiv 1$ so that

$$I = \int_0^\infty f[(cx - a/x)^2]F(x)\, dx. \qquad (6)$$

Introducing here a new variable

$$cx - a/x = y,$$

we find from (6) that

$$I = \frac{1}{2c} \int_{-\infty}^\infty f(y^2)[1 + y(4ac + y^2)^{-1/2}]\, dy;$$

breaking up the range of integration into $(-\infty, 0)$ and $(0, \infty)$ and changing y to $-y$ in the first one, we get finally

$$I = \int_0^\infty f[(cx - a/x)^2]\, dx = \frac{1}{c} \int_0^\infty f(y^2)\, dy.$$

For instance, taking $f(u) = e^{-u}$ we find

$$\int_0^\infty e^{-c^2x^2 - a^2/x^2}\, dx = \frac{\pi^{1/2}}{2c}\, e^{-2ac},$$

and taking $f(u) = 1/(1 + u^2)$, we have

$$\int_0^\infty \frac{x^2\, dx}{c^2x^4 + (1 - 2ac)x^2 + a^2} = \frac{1}{c} \int_0^\infty \frac{dy}{1 + y^2} = \frac{\pi}{2c}.$$

Consider the integral

$$I = \frac{1}{\sqrt{\pi}} \int_{-\infty}^\infty e^{-y^2}f(x + 2y\sqrt{h})\, dy$$

where f is analytic. Recalling the operatorial form of Taylor's theorem

$$f(x + a) = e^{aD}f(x), \quad D \equiv \frac{d}{dx},$$

we have

$$I = \left[\frac{1}{\sqrt{\pi}} \int_{-\infty}^{\infty} e^{-y^2} e^{2yh^{1/2}D} \, dy \right] f(x) = e^{hD^2} f(x)$$

by evaluating the integral in the square brackets as though D were a parameter. Therefore,

$$\frac{1}{\sqrt{\pi}} \int_{-\infty}^{\infty} e^{-y^2} f(x + 2y\sqrt{h}) \, dy = \sum_{n=0}^{\infty} \frac{h^n}{n!} f^{(2n)}(x).$$

For instance, taking

$$f(x) = xe^{-kx^2} \qquad \text{(with } 4hk + 1 > 0\text{)}$$

we find

$$\sum_{n=0}^{\infty} \frac{h^n}{n!} \frac{d^{2n} xe^{-kx^2}}{dx^{2n}} = x(1 + 4hk)^{-3/2} e^{-kx^2/(1+4hk)}.$$

8. INTEGRATION AND MENSURATION

Let $P_1 P_3$ be a straight segment and P_2 a point on it, and let (x_i, y_i) be the coordinates of P_i and θ the angle of inclination of the segment to a fixed line in the plane. Let $|P_1 P_2| = a_3$, $|P_2 P_3| = a_1$, $|P_1 P_3| = -a_2$, so that $a_1 + a_2 + a_3 = 0$. Suppose now that the segment moves in the plane, returning to its original position after making one complete turn. The motion is otherwise arbitrary but we assume that each P_i describes a simple closed curve in the same sense of rotation; let the curve traced out by P_i enclose area A_i. Then, observing the relations between θ, the a_i's, x_i's, and y_i's, we have by taking differentials

$$\sum_{i=1}^{3} \frac{a_i}{2} (x_i \, dy_i - y_i \, dx_i) + \frac{1}{2} a_1 a_2 a_3 \, d\theta = 0. \qquad (1)$$

Since A_i is given by

$$A_i = \frac{1}{2} \oint (x_i \, dy_i - y_i \, dx_i)$$

we have by integrating (1)

$$\sum_{i=1}^{3} a_i A_i + \pi a_1 a_2 a_3 = 0.$$

In particular, if a sufficiently short straight segment of length d slides all the way around inside a closed simple curve enclosing area A, and if a point P on that segment divides it in the ratio $k : 1$, then P traces out a curve enclosing area B, where

$$B = A - \pi k d^2 / (1 + k)^2.$$

In words: A exceeds B by the area of the ellipse whose semi-axes are the two lengths into which P divides the sliding segment. The reader may wish to prove (1), and to investigate various generalizations in 2, 3, or more dimensions.

Let a plane region R enclosed by a simple closed curve C be rotated about a line L in its plane, which does not cross it, thus generating a solid S. Then the Pappus–Guldin theorems state that:

(A) the volume of S is the area of R multiplied by the length of the path traversed by the centroid of R,
(B) the (lateral) area of S is the circumference of C multiplied by the length of the path traversed by the centroid of C.

These theorems are easily provable by elementary integration and are occasionally useful, both directly for the volume and surface computations, and in reverse, for the finding of centroids. However, as observed by E. J. Routh [21], the Pappus–Guldin theorems are capable of considerable generalization: it is not necessary that there should be a fixed axis of revolution but only an instantaneous axis. In the first place, there is no need for a complete revolution in (A) and (B). It is now possible to suppose that a revolution through an angle α_1 about some L_1 is followed by a revolution through an angle α_2 about some L_2 (which need not even be in the same plane with L_1), and so on. With a suitable passage to the limit, we have the following generalizations of (A) and (B). Let K be an arc of a space curve and let, as before, R be a plane region bounded by a simple closed curve C. Suppose that R moves so that its centroid traverses K and its instantaneous plane is at all times orthogonal to the instantaneous tangent to K. Suppose that K is such that no multiple points occur during the motion (i.e., R sweeps through every point in space at most once). Then the volume of the solid S swept out by R is the area of R multiplied by the length of K. Again, if R moves so that the centroid of C stays on K then under the same conditions the lateral area of S is the circumference of C multiplied by the length of K.

9. REDUCTION OF MULTIPLE INTEGRALS TO SINGLE ONES

A ball B of radius a has variable density $f(r)$ which depends only on the distance r from the center O of B. To find the moment of inertia I with respect to an axis through O we place a Cartesian coordinate system with the origin at O; if I_x, I_y, I_z are the moments of inertia with respect to the axes, we find

$$I = I_x = \iiint\limits_B f(\sqrt{x^2 + y^2 + z^2})(y^2 + z^2)\, dx\, dy\, dz,$$

$$I = I_y = \iiint\limits_B f(\sqrt{x^2 + y^2 + z^2})(x^2 + z^2)\, dx\, dy\, dz,$$

$$I = I_z = \iiint\limits_B f(\sqrt{x^2 + y^2 + z^2})(x^2 + y^2)\, dx\, dy\, dz$$

whence by adding the three equations

$$3I = 2 \iiint\limits_B f(\sqrt{x^2 + y^2 + z^2})(x^2 + y^2 + z^2)\, dx\, dy\, dz$$

and passing to spherical coordinates

$$I = \frac{8\pi}{3} \int_0^a r^4 f(r)\, dr.$$

Consider the multiple integral

$$I = \int \cdots \int\limits_R f(x_1 + \cdots + x_n) x_1{}^{a_1 - 1} \cdots x_n{}^{a_n - 1}\, dx_1 \cdots dx_n$$

extended over the n-dimensional simplex

$$R = \left\{ (x_1, \ldots, x_n) : 0 \leq x_i,\ \sum_1^n x_i \leq 1 \right\}.$$

We introduce new variables $u_1, u_2, \ldots, u_n$:

$$x_1 + x_2 + \cdots + x_n = u_1, \qquad x_2 + \cdots + x_n = u_1 u_2, \qquad \ldots, \qquad x_n = u_1 u_2 \cdots u_n$$

so that

$$x_1 = u_1 - x_2 - \cdots - x_n, \qquad x_2 = u_1 u_2 - x_3 - \cdots - x_n, \qquad \ldots, \qquad x_n = u_1 u_2 \cdots u_n$$

and we find that x_i is expressed in terms of $u_1, \ldots, u_i$ and $x_{i+1}, \ldots, x_n$. It can be therefore shown, [21], that the Jacobian assumes the simple form

$$J = \prod_{i=1}^n \frac{\partial x_i}{\partial u_i}$$

so that

$$J = \prod_{i=1}^n u_i{}^{n-i}.$$

The simplex R is mapped by the transformation in a $1 : 1$ fashion onto the cube

$$H = \{(u_1, \ldots, u_n) : 0 \le u_i \le 1, i = 1, \ldots, n\},$$

and the integral becomes

$$I = \frac{\Gamma(a_1)\Gamma(a_2) \cdots \Gamma(a_n)}{\Gamma(a_1 + a_2 + \cdots + a_n)} \int_0^1 f(u)u^{a_1 + \cdots + a_n - 1} \, du.$$

This has some applications to probability: let $n - 1$ points be taken independently and uniformly at random on a unit interval I and let these points divide I into n intervals of length $I_1, \ldots, I_n$; if E denotes mathematical expectation, then employing the integral I above, one can show that

$$E\left(\prod_{i=1}^n I_i^{a_i - 1}\right) = (n - 1)! \prod_{i=1}^n \Gamma(a_i)\Gamma\left(\sum_{i=1}^n a_i\right), \qquad a_i \ge 0.$$

Hence it is possible to find a variety of moments of integral and non-integral order. If B is the ball

$$B = \left\{(x_1, \ldots, x_n) : \sum_1^n x_i^2 \le c^2\right\}$$

and $b^2 = \sum_{i=1}^n a_i^2$, then the integrals

$$I_1 = \int \cdots \int_B F\left(\sum_1^n a_i x_i\right) dx_1 \cdots dx_n,$$

$$I_2 = \int \cdots \int_B F\left(\sum_1^n a_i x_i\right)\left(c^2 - \sum_1^n x_i^2\right)^{-1/2} dx_1 \cdots dx_n$$

reduce, by simple transformations, to single integrals:

$$I_1 = \frac{\pi^{(n-1)/2}}{\Gamma[(n+1)/2]} \int_{-c}^c F(by)(c^2 - y^2)^{(n-1)/2} \, dy,$$

$$I_2 = \frac{\pi^{n/2}}{\Gamma(n/2)} \int_{-c}^c F(by(c^2 - y^2)^{(n/2)-1} \, dy.$$

The n-tuple iterated integral

$$I_n(f) = \int_{x_0}^x \cdots \int_{x_0}^x f(x)dx \cdots dx$$

reduces to a single integral:

$$I_n(f) = \frac{1}{(n-1)!} \int_{x_0}^x (x - y)^{n-1} f(y) \, dy \tag{1}$$

as can be verified in several ways, for instance, by differentiation with respect to a parameter, by integration in two different ways of $f(x)$ over the simplex

$$R = \{(x_1, \ldots, x_n) : x_0 \leq x_1 \leq x_2 \leq \cdots \leq x_n \leq x\},$$

or by using some elementary properties of the Laplace transforms and writing $I_n(f)$ as a multiple convolution. We observe that on replacing $(n - 1)!$ in (1) by $\Gamma(n)$ we have here a possible definition of integration of fractional order n.

10. INTEGRATION AS A SOURCE OF NEW FUNCTIONS

In this section we outline very briefly and superficially the role of integration as a source of mathematical novelty. Integration is one of the principal means whereby we increase our store of functions because, roughly speaking, when $f(x)$ is of some type, then $f'(x)$ is of the same or simpler type whereas $\int f(x)\, dx$ is usually of a more complex type. Briefly: differentiation does not increase functional complexity, integration does. Since integrating, $F(x) = \int f(x)\, dx$, is a special case of solving a differential equation, $F' = f$, it is natural to consider the general problem of deciding when differential equations, involving as coefficients functions of certain simple types, have solutions which are new types of functions. The theory which handles such questions was created in the middle of the nineteenth century by Liouville, and continued by Ostrowski [58] and Ritt [64]. Liouville has built upon some suggestions of Laplace and, possibly, upon the work of Abel on the unsolvability of quintic equations by radicals. This has been proved independently by Ruffini, Abel, and Galois. The proof of Galois proceeds by considering field extensions and groups of substitutions. Abel considered expressions involving radicals and used some ideas concerning what might be called the degree of complexity of such expressions. As observed by Ritt [64], this may have inspired Liouville. Today Abel's proof is practically forgotten for it is Galois' proof that gave rise to the Galois theory of fields and, indirectly, to much of modern algebra and related branches of mathematics. Yet, the work of Liouville and the currently active field of algorithmic complexity, which may be said to have been initiated by Abel's result, testify to the fertility of ideas in Abel's proof.

Joseph Liouville (1809–1882) has introduced into mathematics transcendental numbers, degree of approximation of algebraic numbers, and derivatives of fractional order; he was one of the founders of analytic function theory, the theory of boundary-value problems, the theory of elliptic functions, and statistical mechanics. In geometry he proved the theorem referred to at the end of Section 5, Chapter 1, which describes the conformal group in three dimensions, and thus precludes the existence of a function theory of a "three-dimensional" complex variable. He also determined the class of all surfaces, other than the obvious family of surfaces of revolution, whose geodesics are expressible by integrals [71].

We start our store of functions with x, its nonnegative integer powers, and their linear combinations, the polynomials. These form a differential ring being closed under addition, multiplication, and differentiation. To obtain closure under division we enlarge the class of polynomials to rational functions. The class R of rational functions is a differential field. R is also closed under some less obvious operations: the Hadamard product $f \circ g$ and the idempotency operation $f \to \hat{f}$. Let

$$f(x) = \sum_{n=0}^{\infty} f_n x^n, \qquad g(x) = \sum_{n=0}^{\infty} g_n x^n \tag{1}$$

be rational functions, then

$$(f \circ g)(x) = \sum_{n=0}^{\infty} f_n g_n x^n, \qquad \hat{f}(x) = \sum_{n=0}^{\infty} \operatorname{sgn}|f_n| x^n$$

are also rational. The rationality of $f \circ g$ is a fairly trivial matter (see Section 5, Chapter 6). But the rationality of $\hat{f}$ is equivalent [46] to a rather deep theorem due to Siegel [67], Mahler [43], and Lech [40]. This theorem asserts that if in (1) $f_n = 0$ for infinitely many n then the set $\{n : f_n = 0\}$ consists of a finite (or empty) exceptional set and one or more arithmetic progressions with the same common difference.

In addition to rational functions we acquire early in the game the algebraic functions, usually first those like $\sqrt{x}$ or $\sqrt[3]{2x^2 + 1} + \sqrt{x - 1}$ which are explicitly expressible by radicals, then those like $y = f(x)$ where

$$(3x + 1)y^5 - xy + x^2 + 1 = 0$$

which are not so expressible. The set A of all algebraic functions forms an algebraically closed differential field. However, unlike R, A is not closed under the Hadamard product:

$$f(k) = \sqrt{\frac{\pi}{2}} \sum_{n=0}^{\infty} 2^{-2n} \binom{2n}{n} k^{2n} = \sqrt{\frac{\pi}{2}} (1 - k^2)^{-1/2}$$

is an algebraic function of k but

$$f(k) \circ f(k) = \frac{\pi}{2} \sum_{n=0}^{\infty} \left[2^{-2n} \binom{2n}{n} \right]^2 k^{2n}$$

is the complete elliptic integral

$$K(k) = \int_0^{\pi/2} \frac{d\phi}{\sqrt{1 - k^2 \sin^2 \phi}}$$

which is not algebraic in k. As an aside we observe that if $J_0(x)$ and $I_0(x)$ are the Bessel functions

$$J_0(x) = 1 - \frac{x^2}{2^2} + \frac{x^4}{2^2 \cdot 4^2} - \frac{x^6}{2^2 \cdot 4^2 \cdot 6^2} + \cdots$$

$$I_0(x) = 1 + \frac{x^2}{2^2} + \frac{x^4}{2^2 \cdot 4^2} + \frac{x^6}{2^2 \cdot 4^2 \cdot 6^2} + \cdots$$

then

$$I_0(x) = e^{x^2/2} \circ e^{x^2/2}, \qquad J_0(x) = e^{x^2/2} \circ e^{-x^2/2}.$$

Also, the dilogarithm $L(x)$ of Section 4 may be written as

$$L(x) = [\log(1 + x)] \circ [\log(1 + x)].$$

The reader may wish to investigate possible connections with the Dirichlet operators of Section 9, Chapter 2 and with the Lambert series of Section 14, Chapter 2.

Suppose that we stay in the domain R of rational functions, can we integrate $1/x$? That is, we ask: Is

$$\int_1^x \frac{dx}{x} = f(x)$$

rational? No, as can be seen in a variety of ways. For instance, from an examination of the partial fraction decomposition of f and f' it follows that the equation $f'(x) = 1/x$ is impossible for any rational function f. Alternatively, we can show from the definition of f that it tends to infinity with x, but more slowly than any positive power of x; hence again f cannot be rational. Could $f(x)$ be an algebraic function? No, for much the same reason, though here the details of the proof would be slightly harder. Since an algebraic function has a fractional-power series expansion, it behaves at infinity like a rational power of x; hence again $f(x)$ cannot be algebraic. This form of proof, that $f(x) \equiv g(x)$ is impossible because f and g behave differently at singular points (or at infinity) occurs fairly often in Liouville theory.

Since the integration of $1/x$ takes us outside A we adjoin $f(x)$ to our store of functions and call it by its usual name $\log x$. What else do we get together with it? Observe that the algebraic functions A are closed under inverses: if $f \in A$ then $f^{-1} \in A$. We wish to keep this useful closure property and so, together with $y = \log x$ we also adjoin to our store the inverse, the exponential function $y = e^x$. Since

$$\cos x = (e^{ix} + e^{-ix})/2, \qquad \arcsin x = -i \log(ix + \sqrt{x^2 - 1})$$

we can synthesize all the trigonometric, hyperbolic, and inverse trigonometric or hyperbolic functions out of $\log x$, e^x, and algebraic functions. In fact, simply by adjoining to A the logarithms and exponentials, we obtain the full store E of the standard, or elementary, functions of analysis. The functions in E are classified by the Liouville scheme: $f \in E$ is of order 0 if it is algebraic, $\log f$ and e^f are then of order 1 (assuming $f \neq$ const.) and so is any function algebraic in x, $\log f$, and e^f; $f \in E$ is of order n if it is not of order $<n$ and if it is an algebraic combination of functions of order $<n$ and their logarithms and exponentials.

We outline next the method of Hermite [26], whereby the *rational* part of the integral

$$H(x) = \int \frac{P(x)}{Q(x)} \, dx$$

of a rational function P/Q can be found. Of course, the whole of $H(x)$ can be evaluated but this involves partial fractions and the solution of $Q(x) = 0$. The essence of Hermite's method is that it is constructive; it avoids factorization of polynomials and, in fact, it uses only the algorithm for finding the greatest common divisor (A, B) of two polynomials A and B, and similar elementary algebra. First, $Q(x)$ is represented as

$$Q(x) = \prod_{i=1}^{p} [Q_i(x)]^i$$

where each Q_i is a polynomial with simple roots and no two Q_i's have a common factor. In effect, we have grouped together the factors of $Q(x)$ by like multiplicities. This representation of $Q(x)$ can be found by simple algebra, using the quantities like $(Q, Q^{(k)})$. By further use of the greatest common divisor algorithm we find the representation

$$\frac{P}{Q} = \sum_{i=1}^{p} \frac{A_i}{Q_i^i} \, ;$$

therefore it will suffice to determine the rational part of

$$\int \frac{A(x)}{Q_i^n(x)} \, dx$$

where $(Q_i, Q_i') = 1$. We find polynomials B and C such that

$$BQ_i + CQ_i' = A$$

obtaining

$$\frac{A}{Q_i^n} = \frac{B}{Q_i^{n-1}} + C \frac{Q_i'}{Q_i^n}.$$

Integrating this yields the recursion formula

$$\int \frac{A}{Q_i^n} \, dx = \frac{C}{(1-n)Q_i^{n-1}} + \int \frac{D}{Q_i^{n-1}} \, dx$$

with $D = B + C'(n-1)$. Applying this recursion technique $n-1$ times, we obtain eventually

$$\int \frac{A}{Q_i^n} \, dx = R_i + \int \frac{F}{Q_i} \, dx$$

where $R_i(x)$ is the rational part of the integral. That is, the integration of F/Q_i leads to logarithms and arc-tangents alone. Thus, the rational part of $H(x)$ is the sum $\sum R_i(x)$. In particular, Hermite's algorithm allows us to find by constructive means when a rational function has a rational integral.

Liouville's purpose was to investigate the problem of integration in finite terms: if the integrand $f(x)$ belongs to E or to a certain subclass of E, when is

$$F(x) = \int f(x) \, dx$$

a member of E? Liouville's basic theorem is the following: if $f \in A$ and $F \in E$ then

$$F(x) = f_0(x) + \sum_{i=1}^{k} c_i \log f_i(x) \qquad (2)$$

where the c_i are constants and $f_0, \ldots, f_k$ are algebraic functions. The proof of this proceeds by observing that no exponentials can enter F since $F' = f$ is algebraic while, to put it roughly, exponentials survive differentiation. The same is true about any logarithmic term which enters into F otherwise than linearly. Thus we have the form of (2) for F. Finally, the order of each f_i is shown to be 0 since logarithms of higher orders also "partially" survive.

An alternative proof may be obtained by observing that if a (different) constant is added to each logarithm in (2) then the result remains an integral of f. This is so because the net number of arbitrary constants created in this process is one, and the one constant we absorb into the constant of integration (not shown in (2) explicitly). Conversely, the representation of F must be of such a form that the above economy of arbitrary constants is maintained.

As an application of the above theorem of Liouville it can be shown that the elliptic integral

$$I(x) = \int \frac{dx}{\sqrt{(1-x^2)(1-k^2x^2)}}, \qquad k \neq 0, 1 \qquad (3)$$

is not an elementary function. One further result is needed here: a theorem due to Abel [64] which states that if f is algebraic and its integral F is elementary, then a representation (2) of F can be found, such that each $f_i (i = 0, 1, \ldots, k)$ is rational in x and f.

Before proving that $I(x) \notin E$ we prove first a general result that if $f \in A$ and $F \in E - A$ then f has a singularity with residue $\neq 0$. To show this we suppose that the integer k in (2) is minimal; as a consequence the numbers $c_1, \ldots, c_k$ are linearly independent over the integers. By differentiating (2) we have

$$f(x) = f_0'(x) + \sum_{i=1}^{k} c_i f_i'(x)/f_i(x)$$

and by Abel's theorem $f_0, f_1, \ldots, f_k$ may be taken to be rational in x and f. The residues of $f_0'(x)$ are 0 and our proposition follows now from the linear independence of the c_i's.

Denote the integrand in (3) by $f(x)$. If $I(x)$ were elementary but not algebraic then by the above general result $f(x)$ would have had a singularity with a residue $\neq 0$, which is not the case. If $I(x)$ were algebraic then by Abel's theorem it would have been rational in x and f. This possibility is excluded by a not very hard further analysis of behavior at the singularities, and so it follows that $I(x) \notin E$.

Another general theorem of Liouville is concerned with the integrals of the form

$$F(x) = \int g(x) e^{f(x)} \, dx$$

where f and g are algebraic and $f \neq$ const.; the result is that F, if it is elementary, is of the form

$$F(x) = w(x) e^{f(x)}$$

where w is rational in $x, f,$ and g. As an application, the integrals

$$\int e^{-x^2} \, dx, \qquad \int \frac{e^x}{x} \, dx$$

can be shown to be non-elementary, for otherwise w's would have to be rational functions satisfying respectively the differential equations

$$w' - 2xw = 1, \qquad w' - w/x^2 = 1/x.$$

Now, a simple analysis concerning poles and residues shows that neither equation can have a rational solution.

We mention next a related result of Chebyshev [64]: the binomial integral

$$\int x^p(a + bx^r)^q \, dx,$$

where p, q, r are rational, is an elementary function if and only if at least one of the following three numbers is an integer:

$$(p + 1)/r, \quad q, \quad q + (p + 1)/r.$$

In each case a simple substitution leads to evaluating the integral. As an application the reader may wish to prove the following: let C be an arc of the curve $y = ax^{m/n}$ where m and n are relatively prime integers, let $L(C)$ be the length of C, $A_x(C)$ the area of the surface obtained by rotating C about the x-axis, and $A_y(C)$ the same for the y-axis. Then $L(C)$ is expressible by elementary functions if and only if $2m - 2n | n$ or $2m - 2n | m$; $A_x(C)$ if and only if $2m - 2n | m + n$ or $m - n | m$; $A_y(C)$ if and only if $m - n | n$ or $2m - 2n | m + n$.

The Kepler equation

$$z = a + x \sin z,$$

which arises in astronomy, was mentioned in Section 10, Chapter 3. Liouville showed that if z is expressed as a function of a, $z = f(a)$, then f is not an elementary function. This may also be put as follows: the elementary function $g(z) = z - x \sin z$ does not have an elementary inverse. A necessary and sufficient condition for an elementary function (of order >0) to have an elementary inverse is given in Ritt [63], [64].

A function $y = f(x)$ is said to satisfy an algebraic differential equation if

$$P(x, y, y', \ldots, y^{(n)}) = 0$$

where P is a polynomial with constant coefficients. Every algebraic function y satisfies trivially such an equation, with $n = 0$. Suppose next that y is an elementary function of order $n > 0$. If it involves terms like $u = \log g$ or $v = e^f$, with f or g of order $<n$, then we have $gu' = g'$ or $v' = fv$. Thus successive differentiation eliminates logarithms and exponentials, and we can show by an induction on n that y satisfies an algebraic differential equation. A classical result of Hoelder [32], [29] asserts that the Γ-function does not satisfy any algebraic differential equation. Roughly speaking, the proof of this proceeds by an induction on the order of the hypothetical differential equation. It is shown by an analysis of the behavior at the poles that the algebraic differential equation and the difference equation $\Gamma(x + 1) = x\Gamma(x)$ satisfied by the Γ-function are incompatible. It follows as a corollary that $\Gamma(x)$ is not an elementary function. Using the functional equation of the

Riemann Zeta function, the reader may wish to show that it too is not elementary and does not satisfy an algebraic differential equation.

Finally, in the domain of differential equations Liouville proved some theorems from which it follows, for instance, that the Bessel differential equation

$$x^2 y'' + xy' + (x^2 - n^2)y = 0$$

has elementary solutions if and only if $2n$ is an odd integer [76].

6

MISCELLANEOUS
INTERMEDIATE TOPICS

1. PRINCIPLE OF SPECIALIZATION AND GENERATING FUNCTION

We illustrate this first on the elementary example of the nth derivative of a product of two functions. Successive applications of the Leibniz rule $(uv)' = uv' + u'v$ show that

$$(uv)^{(n)} = \sum_{k=0}^{n} A_{nk} u^{(n-k)} v^{(k)} \tag{1}$$

where the coefficients A_{nk} are independent of the particular functions u and v. Therefore we *may specialize* by choosing u and v suitably. With

$$u(x) = e^{ax}, \qquad v(x) = e^{bx}$$

(1) becomes after some simplification

$$(a + b)^n = \sum_{k=0}^{n} A_{nk} a^{n-k} b^k;$$

regarded as an identity in the indeterminates a and b, this is a *generating function* for the numbers A_{nk}. Hence by the binomial theorem $A_{nk} = \binom{n}{k}$ and (1) becomes

$$(uv)^{(n)} = \sum_{k=0}^{n} \binom{n}{k} u^{(n-k)} v^{(k)}. \tag{2}$$

Similar use of the multinomial theorem

$$(a_1 + a_2 + \cdots + a_j)^n = \sum_{k_1 + k_2 + \cdots + k_j = n} \binom{n}{k_1 \, k_2 \ldots k_j} a_1^{k_1} a_2^{k_2} \cdots a_j^{k_j},$$

213

where

$$\binom{n}{k_1 k_2 \cdots k_j} = n!/(k_1! \, k_2! \cdots k_j!)$$

is the multinomial coefficient, leads to a generalization of (2): if $u_1(x), \ldots,$ $u_j(x)$ are j functions of x then the nth derivative of their product is

$$(u_1 u_2 \cdots u_j)^{(n)} = \sum_{k_1 + k_2 + \cdots + k_j = n} \binom{n}{k_1 k_2 \ldots k_j} u_1^{(k_1)} u_2^{(k_2)} \cdots u_j^{(k_j)}.$$

A less elementary application of the same type occurs in the following proof of the Faa di Bruno formula for the nth derivative $D_x^n f(g)$ of a composite function $f(g) = f[g(x)]$. Using the ordinary chain rule for differentiation, we find successively

$$\begin{aligned}
D_x^1 f(g) &= f'g' \\
D_x^2 f(g) &= f'g'' + f''g'^2 \\
D_x^3 f(g) &= f'g''' + f''(3g'g'') + f'''g'^3 \\
D_x^4 f(g) &= f'g^{(iv)} + f''(4g'g''' + 3g''^2) + f'''(6g'^2 g'') + f^{(iv)} g'^4
\end{aligned} \tag{3}$$

and so on. Here $f^{(k)}$ denotes the kth derivative of $f(g)$ with respect to g. The general case is of the form

$$D_x^n f(g) = \sum_{k=1}^{n} f^{(k)} A_{nk}(g', g'', \ldots, g^{(n)}) \tag{4}$$

where the coefficients

$$A_{nk} = A_{nk}(g', g'', \ldots, g^{(n)})$$

do not depend on f. Therefore we are free to specialize by choosing f to suit our convenience and we choose again f so that the derivatives $f^{(k)}$ are simply formed: $f(u) = e^{au}$. Further, we are free to choose g to be analytic (we shall be using Taylor series). With these choices (4) becomes

$$e^{-ag(x)} D_x^n e^{ag(x)} = \sum_{k=1}^{n} a^k A_{nk}. \tag{5}$$

We multiply (5) by $t^n/n!$ and sum over n from 1 to ∞; recalling the form of the Taylor series for $g(x + t)$ in powers of t, we may write the result as

$$e^{a[g(x+t)-g(x)]} = 1 + \sum_{n=1}^{\infty} \sum_{k=1}^{n} a^k t^n A_{nk}/n!$$

thus obtaining the *generating function* for the expressions A_{nk}. Developing the difference $g(x + t) - g(x)$ in powers of t we have

$$\prod_{j=1}^{\infty} e^{at^j g^{(j)}/j!} = 1 + \sum_{n=1}^{\infty} \sum_{k=1}^{n} a^k t^n A_{nk}/n! \tag{6}$$

and expanding each factor of the infinite product in the exponential series,

$$\prod_{j=1}^{\infty} \sum_{m_j=0}^{\infty} \frac{1}{m_j!} a^{m_j} t^{jm_j} [g^{(j)}/j!]^{m_j} = 1 + \sum_{n=1}^{\infty} \sum_{k=1}^{n} a^k t^n A_{nk}/n!;$$

here the index m_j refers to the jth exponential series. When all the power series are multiplied out the typical factor is

$$a^{m_1+m_2+\cdots+m_s} t^{m_1+2m_2+\cdots+sm_s} (g'/1!)^{m_1} (g''/2!)^{m_2} \cdots (g^{(s)}/s!)^{m_s}/m_1! \cdots m_s!$$

because in multiplying out infinitely many power series, all with the constant term 1, we take a positive integer s, choose s different series, and then select a term from each series chosen. When $s = 0$ the typical factor is just 1. Now we equate the coefficients of $a^k t^n$ on both sides of (6) and we obtain

$$A_{nk} = n! \sum_{m_1} \sum_{m_2} \cdots \sum_{m_s} (g'/1!)^{m_1} (g''/2!)^{m_2} \cdots (g^{(s)}/s!)^{m_s}/m_1! \, m_2! \cdots m_s! \quad (7)$$

where the summation is over all values $m_1, m_2, \ldots, m_s$ which meet the following conditions:

(a) $m_1, \ldots, m_s$ are positive integers,

(b) $m_1 + \cdots + m_s = k$ (to assure the same power a^k),   (8)

(c) $m_1 + 2m_2 + \cdots + sm_s = n$ (to assure the same power t^n).

When A_{nk} from (7) is substituted into (4) we obtain the Faa di Bruno formula.

EXAMPLE. We consider the case $f[g(x)] = e^{-1/x^2}$ with $f(u) = e^u$ and $g(x) = -1/x^2$. As is well known, f is a common source of counter-examples in analysis on account of its property of being infinitely differentiable for all x, including the value $x = 0$, but failing to be analytic at $x = 0$ and so not having a McLaurin series. To compute

$$\frac{d^n e^{-1/x^2}}{dx^n} \quad (9)$$

we may proceed inductively by showing that

$$\frac{d^n e^{-1/x^2}}{dx^n} = x^{-3n} Q_n(x) e^{-1/x^2} \quad (10)$$

where $Q_0(x) = 1$, $Q_1(x) = 2$, $Q_2(x) = 4 - 6x^2$, and generally, $Q_n(x)$ for $n \geq 1$ is a polynomial of degree $2n - 2$ satisfying the recursion

$$Q_{n+1}(x) = (2 - 3nx^2)Q_n(x) + x^3 Q_n'(x). \quad (11)$$

Alternatively, we may apply the Faa di Bruno formula; computing all the necessary derivatives of g, we find (10) again but this time with an explicit form of $Q_n(x)$:

$$Q_n(x) = (-1)^n n! \sum_{k=1}^{n} (-1)^k \left[\sum_{m_1} \cdots \sum_{m_s} 2^{m_1} 3^{m_2} \cdots (s+1)^{m_s} / m_1! \, m_2! \cdots m_s! \right] x^{2n-2k} \tag{12}$$

where the inner summation is subject to (8). Thus, computing the quantity (9) in two different ways, we obtain the solution (12) of the functional-differential equation (11).

The Faa di Bruno formula (4) can be restated operationally in the following form due to V. F. Ivanoff [34]:

$$D_x^n f(g) = \begin{vmatrix} g'D & g''D & g'''D & g^{(iv)}D & \cdots & g^{(n)}D \\ -1 & g'D & 2g''D & 3g'''D & \cdots & \binom{n-1}{1}g^{(n-1)}D \\ 0 & -1 & g'D & 3g''D & \cdots & \binom{n-1}{2}g^{(n-2)}D \\ 0 & 0 & -1 & g'D & \cdots & \binom{n-1}{3}g^{(n-3)}D \\ \vdots & \vdots & & -1 & \cdots & \\ 0 & 0 & 0 & 0 & \cdots -1 & g'D \end{vmatrix} f(g)$$

where $g^{(k)} = d^k g(x)/dx^k$ and $D^j f(g) = d^j f(g)/dg^j$. As an application we consider the Hermite polynomial $H_n(x)$ given by

$$H_n(x) = (-1)^n e^{x^2/2} \, d^n e^{-x^2/2}/dx^n$$

which, like many other orthogonal polynomials, involves the nth derivative of a composite function. Applying Ivanoff's formula we have the following tri-diagonal representation:

$$H_n(x) = \begin{vmatrix} x & 1 & & & \\ 1 & x & 2 & & \\ & 1 & x & 3 & \\ & & \cdots\cdots\cdots & & \\ & & & 1 & x & n-1 \\ & & & & 1 & x \end{vmatrix};$$

all elements in the determinant, other than those on the main diagonal and its two immediate neighbors, are 0 (hence the name tri-diagonal).

Finally, we observe some combinatorial aspects of Faa di Bruno's formula, in particular, its connection with set-partitions. Let S^n be a set with n elements: $S^n = \{1, 2, \ldots, n\}$. A partition of S^n is a representation

$$S^n = S_1 \cup S_2 \cup \cdots \cup S_k, \qquad k \geq 1,$$

in which the subsets S_i of S^n are nonempty and pairwise disjoint. Let B_n be the number of distinct partitions of S^n, then the first four values of B_n are: $B_1 = 1$, $B_2 = 2$, $B_3 = 5$, $B_4 = 15$. For instance, the fifteen different partitions of S^4 are

one part:	$\{1, 2, 3, 4\}$	1
two parts:	$\{1\}, \{2, 3, 4\} + 3$ more	
	$\{1, 2\}, \{3, 4\} + 2$ more	$4 + 3$
three parts:	$\{1\}, \{2\}, \{3, 4\} + 5$ more	6
four parts:	$\{1\}, \{2\}, \{3\}, \{4\}$	1

giving us the total of 15. The reader may notice the correspondence between the above summands 1, $4 + 3$, 6, 1 and the coefficients in the Faa di Bruno formula (3) for $n = 4$. We observe that this correspondence carries over to the general case: B_n is the sum of all the numerical coefficients in $A_{n1}, A_{n2}, \ldots,$ A_{nn}. Next, we specialize again by choosing $g(x) = e^x$ and $a = 1$, and putting $x = 0$. From (7) we have now

$$B_n = \sum_{k=1}^{\infty} A_{nk}(e^x, e^x, \ldots, e^x)\Big|_{x=0}$$

and therefore by (6)

$$e^{e^t - 1} = 1 + \sum_{n=1}^{\infty} B_n t^n/n! \tag{13}$$

giving us the exponential generating function for the numbers B_n. These are known as Bell numbers; for their combinatorial significance see [4], [61], and [5]. The reader may wish to derive the relation (13) by the symbolic method and the techniques of Section 11, Chapter 3, observing that (13) may be written symbolically as

$$e^{Bt} = e^{e^t - 1}.$$

2. IRRATIONALITY PROOFS

We outline here the Lambert–Legendre irrationality proof which uses the partial fraction expansion connected with a different equation, and shows in one sweep the irrationality of e, π, $J_0(1)/J_1(1)$, and other numbers. We observe first that the power series

$$y = f_a(x) = 1 + \frac{x}{1!a} + \frac{x^2}{2!a(a + 1)} + \frac{x^3}{3!a(a + 1)(a + 2)} + \cdots$$

satisfies the differential equation

$$xy'' + ay' = y$$

in virtue of the identity

$$[(a + 1)(a + 2) \cdots (a + n)]^{-1} + n[a(a + 1) \cdots (a + n)]^{-1}$$
$$= [a(a + 1) \cdots (a + n - 1)]^{-1}.$$

Differentiating the differential equation, we have successively

$$xy''' + (a + 1)y'' = y', \qquad xy^{(iv)} + (a + 2)y''' = y''$$

and generally

$$xy^{(n)} + (a + n - 2)y^{(n-1)} = y^{(n-2)}, \qquad n = 2, 3, \ldots.$$

Therefore

$$y^{(n-2)}/y^{(n-1)} = a + n - 2 + x/(y^{(n-1)}/y^{(n)})$$

and hence

$$\frac{y}{y'} = a + \frac{x}{\dfrac{y'}{y''}} = a + \cfrac{x}{a + 1 + \dfrac{y''}{y'''}} = \cdots$$

so that justifying the passage to the limit, we have

$$\frac{y}{y'} = a + \cfrac{x}{a + 1 + \cfrac{x}{a + 2 + \cfrac{x}{\ddots}}} \tag{1}$$

obtaining a continued fraction for y/y'. When a is rational and $\neq 0, -1, -2,$ $\ldots$ while x is rational and $\neq 0$, it follows from the elementary theory of continued fractions that the right-hand side of (1) is a well-defined convergent expression and necessarily irrational (on account of the nontermination of the continued fraction). Hence finally we find that the number $f_a(x)/f_a'(x)$ is irrational if a and x are rational and satisfy the conditions $x \neq 0$, $a \neq 0$, $-1, -2, \ldots$. We note three special cases.

(A) $a = 1/2$. Here $y = \cosh(2x^{1/2})$ and $y' = x^{-1/2}\sinh(2x^{1/2})$. We take $x = p^2/16q^2$ where p and q are positive integers. It follows that

$$y/y' = p[4q\tanh(p/2q)]^{-1}$$

is irrational, hence $\tanh(p/2q)$ is irrational. Since

$$\tanh x = \frac{e^{2x} - 1}{e^{2x} + 1}$$

it follows that $e^{p/q}$ is irrational. Otherwise put, e^r is irrational for every rational $r \neq 0$.

(B) $a = 1/2$, $x = -q^2/4$, q rational and $\neq 0$. Now

$$y/y' = q/2 \tan q,$$

therefore $q^{-1} \tan q$ is irrational if q^2 is rational and $\neq 0$. Suppose now that π is rational, then so is $\pi^2/16$ and therefore by the foregoing the number

$$(\tan \pi/4)/(\pi/4)$$

is irrational; but this number is $4/\pi$, hence π is irrational—a contradiction.

(C) $a = 1$. Here $y = I_0(2x^{1/2})$ where I_0 is the modified Bessel function. Let $x = -p^2/4$ with rational $p \neq 0$; we conclude that $J_0(p)/J_1(p)$ is irrational.

We recall the simple irrationality proof for e: let $n \geq 1$ then

$$e = \sum_{k=0}^{\infty} 1/k! = \sum_{k=0}^{n} 1/k! + \sum_{k=n+1}^{\infty} 1/k! = s_n + r_n$$

say, so that

$$n!e = n!s_n + n!r_n.$$

The first term on the right is an integer and the second one a fraction strictly between 0 and 1 since for $n \geq 1$

$$n! r_n = \frac{1}{n+1}\left[1 + \frac{1}{n+2} + \frac{1}{(n+2)(n+3)} + \cdots\right]$$

which is strictly less than

$$\frac{1}{n+1}[1 + 3^{-1} + 3^{-2} + \cdots] = \frac{4/3}{n+1}.$$

Therefore $n!e$ is never an integer and so ne is never an integer, thus e is irrational. Similar irrationality proofs can be given for

$$J_0(1) = 1 - \frac{1}{2^2} + \frac{1}{2^2 \cdot 4^2} - \frac{1}{2^2 \cdot 4^2 \cdot 6^2} + \cdots, \qquad \sum_{n=0}^{\infty} b_n/n! \text{ (with } b_n = \pm 1)$$

and other similarly formed types of numbers. In this connection the reader may wish to recall the irrationality proofs following Cantor's theorem, in the section on Euler's product, Chapter 3. With more sophisticated, though still elementary, methods we have some theorems of similar type, due to Oppenheim [57]:

(A) let $a_1, a_2, \ldots; n_1, n_2, \ldots; e_1, e_2, \ldots$ be three sequences of positive integers such that

(a) $\lim \sup a_{i+1} n_i^2 n_{i+1}^{-1} \leq 1$,

(b) $a_1 a_2 \cdots a_{i+1}$ g.c.d. $(n_1, n_2, \ldots, n_i)/\text{g.c.d.}(n_1, n_2, \ldots, n_{i+1}) \leq M$,

(c) $1 \leq e \leq E$,

then the series

$$\sum_{i=1}^{\infty} a_1 a_2 \cdots a_i e_i / n_i,$$

if convergent, is rational if and only if i_0 exists such that for $i \geq i_0$

$$n_{i+1} = 1 + a_{i+1} n_i (n_i - 1), \qquad e_i = \text{const.};$$

(B) let $n_1, n_2, \ldots$ be a divergent sequence of positive integers, such that $\lim \sup n_i^2 n_{i+1}^{-1} \leq 1$ and $n_1 n_2 \cdots n_i \leq K n_{i+1}$, then the infinite product

$$\prod_{i=1}^{\infty} (1 + n_i^{-1})$$

is rational if and only if i_0 exists such that $n_{i+1} = n_i^2$ for all $i \geq i_0$.

We finish this section with the following one-sentence proof of the irrationality of $\sqrt{2} : \sqrt{2}$ is irrational since the equation $p^2 = 2q^2$ is impossible as can be seen by examining the last digits of p and q when these two coprime positive integers are written in the ternary scale.

3. PARTITIONS AND EXPANDING INFINITE PRODUCTS IN SERIES

In the previous sections we have already had an occasion to perform such expansions. Here we consider the following problem: evaluate the infinite product

$$P = \prod_{n=1}^{\infty} (1 - 2^{-n}) \tag{1}$$

to 100 digits' accuracy. This is not meant as a joke; the problem, though with lesser accuracy, actually arose in connection with sampling probabilities and, independently, in the theory of linear feedback shift registers. The point is that we shall illustrate the usefulness of converting infinite products into infinite series, and in our case the large accuracy desired forces us to look for fast-convergent series. Since there is no parameter with respect to which to expand in (1) we bring in some additional structure by generalizing (1) to

$$P(x) = \prod_{n=1}^{\infty} (1 - x^n); \tag{2}$$

this will be expanded into a fast-convergent series and we shall then have $P = P(1/2)$. The reader will notice later that introducing a parameter into (1) by letting

$$Q(x) = \prod_{n=1}^{\infty} (1 - x/2^n)$$

and converting the above into a series would not have worked nearly as well. To transform (2) into a power series we use the Euler identity

$$\prod_{n=1}^{\infty} (1 - x^n) = 1 + \sum_{n=1}^{\infty} (-1)^n [x^{n(3n+1)/2} + x^{n(3n-1)/2}]. \tag{3}$$

The genesis of this is in the theory of elliptic functions, which is not important for us here, but there is also a connection of (3) with the theory of partitions, which *is* important. In fact, we shall prove (3) by an argument due to F. Franklin [28] arising from the combinatorics of partitions.

Let A be an infinite sequence $a_1, a_2, \ldots$ of positive increasing integers. By a partition of a positive integer n we mean a representation of n of the form

$$n = b_1 + b_2 + \cdots + b_k \tag{4}$$

where $k \geq 1$, $b_1 \leq b_2 \leq \cdots \leq b_k$, and each b_i is in A. There are two possibilities: either no repetitions are allowed so that each b_i occurs only once in (4) and $b_1 < b_2 < \cdots < b_k$, or repetitions are allowed. For the case without repetitions we have the generating function

$$f(x) = \prod_{n=1}^{\infty} (1 + x^{a_n}). \tag{5}$$

Using the basic expansions

$$\prod_{n=1}^{N} (1 + X_n) = 1 + \sum X_i + \sum\sum X_i X_j + \cdots + X_1 X_2 \cdots X_N$$

and under suitable convergence conditions

$$\prod_{n=1}^{\infty} (1 + X_n) = 1 + \sum_{j=1}^{\infty} \left(\sum_{i_1} \sum_{i_2} \cdots \sum_{i_j} X_{i_1} X_{i_2} \cdots X_{i_j} \right),$$

we find that

$$\prod_{n=1}^{\infty} (1 + x^{a_n}) = 1 + \sum_{n=1}^{\infty} p(n, A, 1)x^n$$

where $p(n, A, 1)$ is the number of ways in which n is representable in the form (4) without repetitions. Similarly, if in the representation (4) each summand is allowed to occur at most $k - 1$ times (where $k > 1$) we find for the number $p(n, A, k - 1)$ of partitions of n the generating function

$$(1 + x^{a_1} + x^{2a_1} + \cdots + x^{(k-1)a_1})(1 + x^{a_2} + x^{2a_2} + \cdots + x^{(k-1)a_2}) \cdots$$

so that

$$\prod_{n=1}^{\infty} \frac{1 - x^{ka_n}}{1 - x^{a_n}} = 1 + \sum_{n=1}^{\infty} p(n, A, k - 1)x^n.$$

In the limiting case of $k \to \infty$, unrestricted repetition is allowed and we have then

$$\left[\prod_{n=1}^{\infty} (1 - x^{a_n}) \right]^{-1} = 1 + \sum_{n=1}^{\infty} p(n, A, \infty)x^n.$$

Other generating functions may be considered, for instance

$$\prod_{n=1}^{\infty} (1 - x^{a_n});$$

here we deal with partitions without repetition and in expanding the product in a power series we have to distinguish between partitions into an even, and those into an odd, number if parts. Let their numbers be pe(n) and po(n), then we have

$$\prod_{n=1}^{\infty} (1 - x^{a_n}) = 1 + \sum_{n=1}^{\infty} [\mathrm{pe}(n) - \mathrm{po}(n)]x^n. \tag{6}$$

The most interesting case arises when the basic sequence A of summands $a_1, a_2, \ldots$ is $1, 2, \ldots$. We have then by (2) and (6)

$$P(x) = 1 + \sum_{n=1}^{\infty} [\mathrm{pe}(n) - \mathrm{po}(n)]x^n.$$

The Euler identity (3) is therefore equivalent to

$$\begin{aligned} \mathrm{pe}(n) - \mathrm{po}(n) &= 0 & n \neq k(3k \pm 1)/2 \\ \mathrm{pe}(n) - \mathrm{po}(n) &= (-1)^k & n = k(3k \pm 1)/2. \end{aligned} \tag{7}$$

In proving (7) we use the standard combinatorial principle of correspondence: since the difference pe(n) $-$ po(n) is of interest we establish, *insofar as possible*, 1:1 correspondence between the partitions into an even number of parts and those into an odd number of parts, canceling the ones against the others and concentrating on those cases when there is no such correspondence. For this purpose we use the Ferrers graph; a partition such as $14 = 5 + 4 + 3 + 2$ is represented as the array of dots of Figure 1a. The dots in the lowest layer are called the base and those on the $45°$ slope line from the rightmost dot are called the slope; the same names will be used for the corresponding numbers of dots. If the base is less than the slope, as in Figure 1a, we move it over to form a

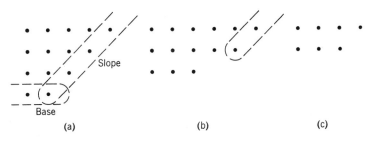

FIGURE 1.   Ferrers graphs.

new slope as in Figure 1b and we establish thereby a $1:1$ correspondence between a partition into an even number of parts: $14 = 5 + 4 + 3 + 2$, and a partition into an odd number: $14 = 6 + 5 + 3$. If the base is equal to the slope we can still perform this unless the base and the slope have a dot in common (exceptional case 1). If the slope is less than the base we reverse the operation and transform Figure 1b into Figure 1a by moving the slope so as to form a new base. This is possible in all cases except for that illustrated in Figure 1c when the base exceeds the slope by one dot which is common to both (exceptional case 2).

Therefore there is a $1:1$ correspondence between the partitions into an even and into an odd number of parts, save for the two exceptional cases. In the first case n is necessarily of the form $k + (k + 1) + \cdots + (2k - 1)$ or $n = k(3k - 1)/2$. In the second case n must be of the form $(k + 1) + (k + 2) + \cdots + (k + k)$ or $n = k(3k + 1)/2$. In either case there is exactly one partition left over from the $1:1$ correspondence, and it is into an even number of parts if k is even, and into an odd number of parts if k is odd. Therefore (7), and so (3), is proved.

We return now to our original problem. From (1), (2), and (3) we have

$$P = 1 + \sum_{n=1}^{\infty} (-1)^n [2^{-n(3n+1)/2} + 2^{-n(3n-1)/2}]$$

which is a very fast-convergent alternating series. To evaluate P to the accuracy of 100 digits we therefore stop just short of the lowest n for which

$$2^{-n(3n-1)/2} + 2^{-n(3n+1)/2} < 10^{-100}.$$

Accordingly, the approximation

$$P = 1 + \sum_{n=1}^{14} (-1)^n [2^{-n(3n-1)/2} + 2^{-n(3n+1)/2}]$$

gives us the desired accuracy.

4. PRINCIPLE OF INFINITE CROWDING

This may be used to advantage in certain problems where one has to show that some metrically defined quantity is infinite or where the existence of an infinity of objects of a certain kind is to be demonstrated. A trivial case of the first type is: if infinitely many segments are subsets of $[0, 1]$ and if the sum of their length is infinite, then some point of the interval $[0, 1]$ is covered by infinitely many of the segments. An even simpler example of the second kind is: if infinitely many objects are distributed between a finite number of locations, then at least one location contains infinitely many objects.

For our first application we prove the Mergelyan–Wesler theorem: let D be a circular disk in the plane and Let $D_1, D_2, \ldots$ be an infinite sequence of circular disks of radii $r_1, r_2, \ldots$, such that: (a) each D_i is a subset of D, (b) any two disks D_i and D_j have at most one point in common, (c) the area of D is the sum of the areas of the D_i's; then

$$\sum_{i=1}^{\infty} r_i = \infty.$$

Consider the circular rims C_1, C_2, of the disks $D_1, D_2, \ldots$ and project them onto some diameter I of D. Then

$$\sum_{i=1}^{\infty} r_i = \frac{1}{2} \sum_{i=1}^{\infty} \text{length}(I_i)$$

where I_i is the projection of C_i onto I. Therefore, to prove our theorem we have to show that the sum $\sum \text{length}(I_i)$ is infinite. This will follow from the fact that the segments $I_1, I_2, \ldots$ form an infinite crowding on I in a very strong sense: almost every point of I belongs to infinitely many of them. To prove this we require the elements of Lebesgue measure theory. For every $x \in I$ let $L(x)$ be the chord of D passing through x and perpendicular to I. Let E be the subset of D left uncovered by $D_1, D_2, \ldots$:

$$E = D - \bigcup_{i=1}^{\infty} D_i.$$

Then E is an uncountable set of plane measure 0. Let A be the subset of I consisting of all points x such that $L(x) \cap E$ has positive linear measure. Then A itself has linear measure 0 by Fubini's theorem since E has plane measure 0. Let $x \in I - A$, then $L(x)$ is completely (in the measure sense) covered by the disks $D_1, D_2, \ldots$; it follows that if it is covered by a finite number of them then it must pass through a point common to two of the disks. But the number of such points is countable (or finite). Hence the set B of exceptional points of this type is at most countable. It follows that every $x \in I - A - B$ belongs to infinitely many I_i's, and so the theorem is proved.

The reader may wish to prove that E is uncountable and to find whether a diameter I exists for which the set A is empty (or finite, or countable); also, to give an example of I for which A is in fact uncountable. The whole theorem may be generalized to higher dimensions and to sets other than disks (in which case the set-diameter replaces the radius). As in many other cases, the reader ought to ask himself whether this theorem represents " the whole truth." It does not, for not only is the sum $\sum r_i$ divergent but it is very likely that every higher moment sum

$$\sum_{i=1}^{\infty} r_i^a$$

is divergent whenever $a < S$ where S is a universal constant whose value is approximately 1.306951 (for the origin of this estimate see [47]). This as yet unproved hypothesis rests on another one, which may be loosely formulated as follows: to make the sum $M_a = \sum r_i^a$ smallest we make the radii r_i largest possible; hence M_a will assume its lowest possible value if from some place i_0 onward each successive disk D_{i+1} is the biggest one fitting into that part of D which is left uncovered by $D_1, D_2, \ldots, D_i$. Under this condition we shall speak of an osculatory packing of D by $D_1, D_2, \ldots$. The previously quoted number 1.306951 arose from numerical experimentation with the first 20,000 disks of an osculatory packing.

We sketch now the proof that for an osculatory packing M_a diverges if

(E) $\qquad a < (\log 3)/[\log(4 + 13^{1/2}) - \log 3] = 1.18096 \ldots$

(considerably better estimates are known now [7], [8]). Let A, B, C be three pairwise externally tangent circles of curvatures a, b, c (we recall that the curvature of a circle is the reciprocal of its radius). Into the concave curvilinear triangle T bounded by the arcs of A, B, and C we inscribe the largest possible circle F, of curvature f, and we call it the 0-th order circle. Similarly, we inscribe into the parts of T uncovered by F three largest possible circles, G, H, K of curvatures g, h, k. These are the three circles of order 1 and we suppose that F separates B from K, C from G, and A from H. The process of successive inscription continues, to give us $N = 3^n$ circles of order n. The Soddy formula, mentioned in Section 9 of Chapter 2 may be applied four times to give us

$$a^2 + b^2 + c^2 + f^2 = (a + b + c + f)^2/2,$$
$$a^2 + b^2 + f^2 + g^2 = (a + b + f + g)^2/2,$$
$$b^2 + c^2 + f^2 + h^2 = (b + c + f + h)^2/2,$$
$$a^2 + c^2 + f^2 + k^2 = (a + c + f + k)^2/2.$$

Subtracting the second, the third, and the fourth equation from the first one, simplifying, and adding, we get

$$h + k + g = 3(a + b + c) + 6f.$$

In the same way we obtain the sum of the nine curvatures of the nine circles of order 2:

$$\sum_{m=1}^{9} c_{m2} = 24(a + b + c) + 45f,$$

and generally for the 3^n nth order circles we have

$$\sum_{m=1}^{3^n} c_{mn} = p(n)(a + b + c) + q(n)f.$$

Recursion formulas are now verified:

$$p(n + 1) = 2p(n) + 3q(n), \qquad q(n + 1) = 3p(n) + 6q(n)$$

and we solve for $p(n)$ and $q(n)$:

$$p(n) = K_1 b_1{}^n + K_2 b_2{}^n, \qquad q(n) = K_3 n_1{}^n + K_4 b_2{}^n$$

where $b_1 = 4 + 13^{1/2}$, $b_2 = 4 - 13^{1/2}$, $K_1 = -K_2 = 3(13)^{1/2}/26$, $K_3 = (13 + 2(13)^{1/2})/26$, $K_4 = (13 - 2(13)^{1/2})/26$. Hence we obtain our principal estimate

$$\sum_{m=1}^{3^n} c_{mn} \sim K b_1{}^n, \qquad K \text{ constant, } n \text{ large.}$$

Next, for the sum of the ath powers of the radii of $D_1, D_2, \ldots$ we have

$$M_a \geq \sum_{n=1}^{\infty} \left(\sum_{m=1}^{3^n} c_{mn}^{-a} \right);$$

therefore by the principal estimate

$$M_a \geq K^{-a} \sum_{n=1}^{\infty} 3^n (b_1/3)^{-na}$$

so that M_a diverges if

$$a < (\log 3)/(\log b_1 - \log 3)$$

thus proving (E).

A second and seemingly quite different application of the infinite crowding principle is the following. Let $D_1, D_2, \ldots$ be an infinite sequence of disjoint disks in the plane, which exhausts some positive fraction of the area of the whole plane. By this we mean the following: let $D(p, R)$ be the circular disk of radius R centered at p and let $f(p, R)$ be the fraction of the area of $D(p, R)$ covered by those disks D_i which are inside $D(p, R)$; then for some (and therefore for every) p we have

$$f(p, R) \geq c, \quad c \text{ a positive constant, } R \text{ large.}$$

Under these conditions we are required to show that from every point p of the plane a ray can be drawn which intersects infinitely many disks D_i. Somewhat imprecisely put, the problem is: if a positive fraction of the area of the plane is occupied by a cloud of two-dimensional circular drops then from every point of the plane there is a direction of sighting occulted by infinitely many drops.

Since p is arbitrary we take it to be the origin and we suppose that it does not lie inside or on any of the disks D_i (for the removal, if necessary, of a disk or a finite number of disks does not really change anything). We consider the angular region A_i consisting of all rays from the origin which pass through the disk D_i and we apply the infinite crowding principle to A_i's and to the corresponding angles α_i: it will be shown that $\sum_{i=1}^{\infty} \alpha_i$ is divergent whence it follows that some ray from the origin cuts infinitely many disks D_i.

Let r_i be the radius of D_i and let d_i be the distance from the origin to the center of D_i. By renumbering, if necessary, we may suppose that $d_1 \leq d_2 \leq \cdots$. We have

$$\sum_1^N \alpha_i = 2 \sum_1^N \text{arc sin } r_i/d_i > \sum_1^n r_i/d_i.$$

Our object now is to show that the series $\sum_1^{\infty} r_i/d_i$ diverges. It may be supposed without loss of generality that

$$\lim_{i \to \infty} r_i/d_i = 0 \tag{1}$$

for otherwise there is nothing to prove. Consider the disk $D(p, d_N)$ and let

$$\rho_N = \max(r_1, r_2, \ldots, r_N);$$

since $d_1 \leq d_2 \leq \cdots$ the disks $D_1, D_2, \ldots, D_N$ lie all inside the disk $D(p, d_N + \rho_N)$. We take N large enough to apply our information on the area:

$$\pi(d_N + \rho_N)^2 \leq c^{-1} \pi \sum_1^N r_i^2$$

with some positive c. It follows that

$$d_N + \rho_N \leq c^{-1/2} \sum_1^N r_i. \tag{2}$$

From (1) it follows that for sufficiently large N

$$\rho_N/d_N \leq (2c^{1/2})^{-1}$$

which together with (2) shows that

$$d_N \le k_1 \sum_1^N r_i$$

for all sufficiently large N, with some positive k. Hence also

$$d_i \le k \sum_{j=1}^{i} r_j \tag{3}$$

for all i, with some $k > 0$. We are now ready to prove that $\sum r_i/d_i$ diverges. First, it is clear that $\sum r_i$ diverges for otherwise the area condition on the disks could not hold. Next, by (3)

$$\sum_{i=1}^{\infty} r_i/d_i \ge k^{-1} \sum_{i=1}^{\infty} \frac{r_i}{\sum_{j=1}^{i} r_j}. \tag{4}$$

We now apply the Abel–Dini theorem: if $r_1 + r_2 + \cdots$ is a divergent series of positive terms and if s_n is its nth partial sum then the series

$$\sum_{i=1}^{\infty} r_i s_i^{-a}$$

converges for $a > 1$ and diverges for $a \le 1$. Applying this to (4) we find that $\sum r_i/d_i$ indeed diverges, and the problem is finished.

As a matter of fact, for the special case $a = 1$ of the Abel–Dini theorem we have

$$\sum_{j=1}^{n} r_j/s_j \sim \log s_n \qquad n \text{ large} \tag{5}$$

which is more than we need. To prove (5) we suppose without loss of generality that $r_n/s_n \to 0$ so that

$$\frac{r_n/s_n}{\log\{1/[1 - (r_n/s_n)]\}} = \frac{r_n/s_n}{\log(s_n/s_{n-1})} \to 1 \tag{6}$$

since

$$\lim_{x \to 0} \frac{x}{\log[1/(1 - x)]} = 1.$$

From (6) it follows that also

$$\left(\sum_{j=1}^{n} (r_j/s_j)\right) \bigg/ \left(\sum_{j=1}^{n} \log(s_j/s_{j-1})\right) \to 1$$

where $s_0 = 1$ by definition. But

$$\sum_{j=1}^{n} \log s_j/s_{j-1} = \log s_n$$

and hence

$$\sum_{j=1}^{n} r_j/s_j \sim \log s_n, \qquad \text{q.e.d.}$$

The reader may wish to find out whether there must be more than one such direction of infinite occultation; also, what reasonable homogeneity-in-the-large conditions on the disks D_i will force the set of infinitely occulting rays to be (a) infinite, (b) dense in the plane, (c) uncountable, (d) of positive measure. The problem may be generalized to higher dimensions, to regions other than circular disks, and in other ways. Finally, the reader may wish to attempt an alternative proof of our theorem by the following, seemingly promising, line of reasoning: let $R(\alpha)$ be the ray from the origin making angle α with the x-axis, assuming the conclusion of the theorem were false it would have followed that on each $R(\alpha)$ there is a "last" point of intersection with the disks D_i, which seems to contradict the area condition.

5. APPLICATIONS OF CERTAIN SPECIAL FUNCTIONS

Legendre polynomials and Bessel functions arise often in problems from mathematical physics which have spherical or cylindrical symmetry. For instance, if alternating current of frequency ω flows through a solid cylindrical conductor C of radius a, and if $j(r)$ is the current density at the distance r from the axis of C, then by applying Maxwell equations we find that $j(r)$ satisfies the modified Bessel equation. Hence $j(r)$ is expressible in terms of the modified Bessel functions which, like the hyperbolic functions and unlike both the trigonometric and the ordinary Bessel functions, are monotone and fast increasing with increasing argument. It follows that the higher the frequency ω, the more constrained the current is to a thin outer layer of C. On account of this so-called skin effect the central core of C is best left out altogether, and one has then a thin hollow cylinder as a more efficient way of propagating high frequencies. In this section we consider the appearance of Legendre polynomials and Bessel functions from the rather unexpected direction of number theory.

Legendre Polynomials.   The zeta function of Riemann

$$\zeta(s) = \sum_{n=1}^{\infty} n^{-s} \tag{1}$$

occurs in number theory on account of its alternative representation (the so-called Euler product) over all primes:

$$\zeta(s) = \prod_p (1 - p^{-s})^{-1}. \tag{2}$$

We obtain the identity of (1) and (2) by developing $(1 - p^{-s})^{-1}$ in an infinite series:

$$\zeta(s) = \prod_p \sum_{m=0}^{\infty} p^{-ms};$$

since each integer is a unique prime-power product we have

$$\sum_{n=1}^{\infty} n^{-s} = \prod_p \sum_{m=0}^{\infty} p^{-ms}.$$

Similarly, we consider other Dirichlet series, i.e., series of the form

$$\phi(s) = \sum_{n=1}^{\infty} a_n n^{-s};$$

such a series may also have an Euler product;

$$\phi(s) = \prod_p f(p^{-s}).$$

If

$$\psi(s) = \sum_{n=1}^{\infty} b_n n^{-s} = \prod_p g(p^{-s})$$

is another Dirichlet series with an Euler product then

$$\phi(s)\psi(s) = \sum_{n=1}^{\infty} \left(\sum_{ij=n} a_i b_j \right) n^{-s} = \prod_p [f(p^{-s})g(p^{-s})].$$

More generally, if we have k Dirichlet series with Euler products,

$$\phi_i(s) = \sum_{n=1}^{\infty} a_{ni} n^{-s} = \prod_p f_i(p^{-s}), \qquad i = 1, 2, \ldots, k,$$

then

$$\prod_{i=1}^{k} \phi_i(s) = \sum_{n=1}^{\infty} \left(\sum_{n_1 n_2 \cdots n_k = n} a_{n_1 1} a_{n_2 2} \cdots a_{n_k k} \right) n^{-s} = \prod_p \left[\prod_{i=1}^{k} f_i(p^{-s}) \right].$$

As a special case let $\phi_1(s) = \phi_2(s) = \cdots = \phi_k(s) = \zeta(s)$, then

$$\sum_{n=1}^{\infty} d_k(n) n^{-s} = \prod_p (1 - p^{-s})^{-k} = \zeta^k(s) \tag{3}$$

where $d_k(n)$ is the number of representations of n as a product of k factors, products differing in the order of factors being counted as distinct.

In investigating the behavior of $\zeta(s)$, Titchmarsh [72] was led to consider the Dirichlet series

$$F(s) = \sum_{n=1}^{\infty} d_k^2(n) n^{-s};$$

we ask: Does $F(s)$ have an Euler product, and if so, what is it? From (3) by the binomial theorem we have

$$\sum_{n=1}^{\infty} d_k(n) n^{-s} = \prod_p \sum_{m=0}^{\infty} \binom{k+m-1}{m} p^{-ms}. \tag{4}$$

Further, if

$$\sum_{n=1}^{\infty} a_n n^{-s} = \prod_p \sum_{m=0}^{\infty} A_m p^{-ms}, \qquad \sum_{n=1}^{\infty} b_n n^{-s} = \prod_p \sum_{m=0}^{\infty} B_m p^{-ms}$$

then

$$\sum_{n=1}^{\infty} a_n b_n n^{-s} = \prod_p \sum_{m=0}^{\infty} A_m B_m p^{-ms},$$

because $a_n = A_{m_1} A_{m_2} \cdots A_{m_k}$ if the integer n has the factorization of the form $n = p_1^{m_1} p_2^{m_2} \cdots p_k^{m_k}$. Therefore from (4) we have

$$F(s) = \sum_{n=1}^{\infty} d_k^2(n) n^{-s} = \prod_p \sum_{m=0}^{\infty} \binom{k+m-1}{m}^2 p^{-ms}.$$

If we put

$$f_k(x) = \sum_{m=0}^{\infty} \binom{k+m-1}{m}^2 x^m$$

then the above may be written as

$$\sum_{n=1}^{\infty} d_k^2(n) n^{-s} = \prod_p f_k(p^{-s}). \tag{5}$$

We now compute $f_k(x)$ explicitly. Let

$$f(x) = \sum_{0}^{\infty} a_n x^n, \qquad g(x) = \sum_{0}^{\infty} b_n x^n$$

be two power series; then their Hadamard product $f \circ g$ is defined as

$$f \circ g(x) = \sum_{0}^{\infty} a_n b_n x^n$$

and we have the integral representation

$$f \circ g(x) = \frac{1}{2\pi i} \int_C f(z) g(x/z)\, dz/z \tag{6}$$

which holds for suitable contours C. In our case we take

$$f(x) = g(x) = \sum_{m=0}^{\infty} \binom{k + m - 1}{m} x^m = (1 - x)^{-k}$$

and we have then $f_k(x) = f \circ f(x)$ so that

$$f_k(x) = \frac{1}{2\pi i} \int_C (1 - z)^{-k} (1 - x/z)^{-k} \, dz/z.$$

Taking here for C the circle of radius $x^{1/2}$ about the origin, and writing $z = x^{1/2} e^{i\theta}$, we obtain

$$f_k(x) = \frac{1}{\pi} \int_0^{\pi} [1 - 2x^{1/2} \cos \theta + x]^{-k} \, d\theta. \tag{7}$$

This may be compared with the Laplace integral representation of the nth Legendre polynomial:

$$P_n(x) = \frac{1}{\pi} \int_0^{\pi} [x - (x^2 - 1)^{1/2} \cos \theta]^{-n-1} \, d\theta$$

and we find that

$$f_k(x) = (1 - x)^{-k} P_{k-1}\left(\frac{1 + x}{1 - x}\right).$$

Therefore (5) becomes

$$\sum_{n=1}^{\infty} d_k^2(n) n^{-s} = \zeta^k(s) \prod_p P_{k-1}\left(\frac{p^s + 1}{p^s - 1}\right).$$

We add a few further remarks on some uses of the Hadamard product $f \circ g$ of power series. First, it is clear that Hadamard multiplication is commutative, associative, and distributive with respect to ordinary addition. Next, we state four well-known equivalent conditions for the power series

$$f(x) = \sum_0^{\infty} a_n x^n$$

to represent a rational function:

  (a) $f(x)$ is a quotient of two polynomials;
  (b) $f(x)$ is a sum of partial fractions

$$f(x) = P(x) + \sum_{k=1}^{N} \sum_{j=1}^{M} A_{jk}(\alpha_k - x)^{-j},$$

where P is a polynomial, A_{jk} and α_k are constants;

(c) there is n_0 such that for $n \geq n_0$ the coefficients a_n satisfy a linear recurrence with constant coefficients:

$$a_{n+1+k} = \sum_{j=0}^{k} c_j a_{n+j}, \qquad n \geq n_0;$$

(d) there is n_0 such that a_n is an exponential polynomial in n for $n \geq n_0$:

$$a_n = \sum_{j=1}^{k} P_j(n) b_j^n, \qquad n \geq n_0,$$

where P_j are polynomials and b_j are constants. The reason for the appearance of n_0 in (c) and (d) is that if a power series represents a rational function f then we may change the first few coefficients and the new series still represents a rational function; in effect, we have added a polynomial to f. Suppppose now that we consider a sequence of numbers $a_0, a_1, \ldots$ whose generating function

$$f(x) = \sum_0^{\infty} a_n x^n$$

is rational. If $E(x)$ is an exponential polynomial and $P(x)$ a polynomial then $P[E(x)]$ is also exponential polynomial. Hence it follows by our condition (d) that the sequence

$$P(a_0), P(a_1), P(a_2), \ldots$$

has as its generating function

$$F(x) = \sum_{n=0}^{\infty} P(a_n) x^n$$

which is also rational. Moreover, if

$$P(x) = \sum_{j=0}^{N} A_j x^j$$

then by using the Hadamard product we find that

$$F(x) = \sum_{j=0}^{N} A_j [f(x) \circ f(x) \circ \cdots \circ f(x)]$$

with j appearances of f in the square brackets, the coefficient of A_0 being the function

$$e(x) = \sum_0^{\infty} x^n = \frac{1}{1-x}$$

which acts as identity for the Hadamard product: $e \circ f = f \circ e = f$ for every f. Using our condition (c), we can also express the above as follows: if $P(x)$ is a polynomial and the sequence a_0, a_1, ... satisfies a linear recurrence with constant coefficients from some n_0 onward, then so does the sequence $P(a_0)$, $P(a_1)$, ...

As an example we consider the sequence f_0, f_1, ... of Fibonacci numbers, defined by

$$f_0 = f_1 = 1, \qquad f_{n+2} = f_n + f_{n+1}, \qquad n = 0, 1, \ldots.$$

The generating function $f(x)$ must be rational by condition (c) and we find it easily:

$$f(x) = (1 + 2x)(1 - x - x^2)^{-1}.$$

It follows that the sequence of the squares of Fibonacci numbers f_0^2, f_1^2, ... also has a rational generating function $F(x)$, given by $F(x) = f \circ f(x)$. To compute F explicitly we use the partial fractions of condition (b):

$$f(x) = (a_1 - x)^{-1} + (a_2 - x)^{-1},$$

$$a_1 = (-1 + 5^{1/2})/2, \qquad a_2 = (-1 - 5^{1/2})/2.$$

For simple partial fractions we have the Hadamard product:

$$\frac{A}{\alpha - x} \circ \frac{B}{\beta - x} = \frac{AB}{\alpha\beta - x}$$

and therefore $F = f \circ f$ gives us

$$F(x) = (a_1^2 - x)^{-1} + 2(a_1 a_2 - x)^{-1} + (a_2^2 - x)^{-1};$$

carrying out some simple algebra we find that

$$F(x) = (1 + 7x - 4x^2)/(x^3 - 2x^2 - 2x + 1).$$

Referring again to condition (c) we find that the squares of the Fibonacci numbers satisfy the recurrence

$$f_{n+3}^2 = 2f_{n+2}^2 + 2f_{n+1}^2 - f_n^2, \qquad n \geq 0.$$

Bessel Functions. In certain number-theoretic problems it is important to estimate possibly precisely the number of integer lattice points in a plane region C. When C is the closed disk of radius $x^{1/2}$ about $(0, 0)$ we have the problem of estimating the number $A(x)$ of integer-valued solutions (u, v) of the inequality $u^2 + v^2 \leq x$. To estimate $A(x)$ we consider, after Landau [39], the Poisson summation formula of Section 12, Chapter 3, but in its finite-sum variant: if $f(x)$ is sufficiently differentiable for $a \leq x \leq \beta$, if $f(\alpha) = f(\beta) = 0$, and if the necessary integrals exist, then

$$\sum_{\alpha \leq n \leq \beta} f(n) = \sum_{k=-\infty}^{\infty} \int_\alpha^\beta f(y) \cos 2\pi k y \, dy. \tag{8}$$

We use this first for $f(y) = x - u^2 - y^2$ with $x > 0$, $-\sqrt{x} \le u \le \sqrt{x}$, $\alpha = -\sqrt{x - u^2}$, $\beta = \sqrt{x - u^2}$, and we have

$$\sum_{-\sqrt{x-u^2} \le n \le \sqrt{x-u^2}} (x - u^2 - n^2) = \sum_{k=-\infty}^{\infty} \int_{-\sqrt{x-u^2}}^{\sqrt{x-u^2}} (x - u^2 - y^2)\cos 2\pi k y \, dy$$

$= \phi(x, u)$, say. Next, we use (8) again, with $f(u) = \phi(x, u)$, $x > 0$, $\alpha = -\sqrt{x}$, $\beta = \sqrt{x}$, and we have

$$\sum_{-\sqrt{x} \le u \le \sqrt{x}} \sum_{-\sqrt{x-u^2} \le n \le \sqrt{x-u^2}} (x - u^2 - n^2) = \sum_{b=-\infty}^{\infty} \int_{-\sqrt{x}}^{\sqrt{x}} \phi(x, u)\cos 2\pi b u \, du.$$

$$(9)$$

The left-hand side of (9) is

$$\sum_{m^2+n^2 \le x} (x - m^2 - n^2) = \sum_{m^2+n^2 \le x} \int_{m^2+n^2}^{x} dy = \int_0^x \sum_{m^2+n^2 \le y} 1 \, dy = \int_0^x A(y) \, dy$$

and the right-hand side is

$$\sum_{j=-\infty}^{\infty} \sum_{b=-\infty}^{\infty} Q(x, k, b)$$

where $Q(x, k, b)$ is given by

$$Q(x, k, b) = \iint_{u^2+v^2 \le x} (x - u^2 - v^2)\cos 2\pi k u \cos 2\pi b v \, du \, dv.$$

Next, we have $Q(x, 0, 0) = \pi x^2/2$ and for $k^2 + b^2 > 0$

$$Q(x, k, b) = x[\pi(k^2 + b^2)]^{-1} J_2[2\pi(k^2 + b^2)^{1/2} x^{1/2}]$$

where $J_2(u)$ is the Bessel function of order 2:

$$J_2(u) = \sum_{n=0}^{\infty} \frac{(-1)^n}{n!(n + 2)!} (u/2)^{2n+2}.$$

Summing over all the solutions of $k^2 + b^2 = n$, we let $U(n)$ be their number and we obtain

$$\int_0^x A(y) \, dy = \pi x^2/2 + \frac{x}{\pi} \sum_{n=1}^{\infty} n^{-1} U(n) J_2[2\pi(nx)^{1/2}].$$

$$(10)$$

By a formal differentiation of (10) with respect to x (which, however, has to be justified) we obtain the Hardy identity

$$\pi x + x^{1/2} \sum_{n=1}^{\infty} n^{-1/2} U(n) J_1[2\pi(nx)^{1/2}] = \begin{array}{ll} A(x) & x \text{ noninteger} \\ A(x) - U(x)/2 & x \text{ integer}. \end{array}$$

It follows from this, by some further estimations, that for x large

$$A(x) = \pi x + 0(x^t), \qquad 1/4 \le t \le 1/3.$$

6. CONTINUATION PRINCIPLE

(a) Introduction. This is used to obtain a complete analytic or geometric entity which is initially given only locally and then extended by some continuation method. The application to analytic functions by the power-series continuation is well known and standard. Similarly, to solve an initial value problem $A(f) = 0, f(0) = y_0$, where A is an operator and $y = f(x)$ is a function of a real variable, we may first obtain a local solution $f(x)$ valid for $0 \leq x \leq a_1$, then continue by solving the initial value problem $A(f) = 0$, $f(a_1) = y(a_1)$, to obtain the continuation $y = f(x)$ valid for $a_1 \leq x \leq a_2$, and so on. Also, the theory of summability may be regarded as a continuation of the sum function in various ways, from the set of convergent series to a wider class of series.

In each case there is a consistency condition or a consistency theorem, which requires or states that the continuation is essentially unique. For analytic functions this is the monodromy theorem which states that a function analytic in a simply connected domain may be continued from any initial point to any other point via any continuation path, with the same final result. For the case of initial value problems we have the semigroup property: $T_{u+v} = T_u T_v$ where $y(x) = T_x y(0)$. In summability we have the requirement of regularity that any summability method must sum a convergent series to the value equal to its ordinary sum; further, two summability methods are consistent if they assign the same sum to each series to which they both apply.

(b) A General Method of Geometric Continuation. Let I and F be two closed disjoint regions in the plane; I stands for "initial" and F for "forbidden". Let P be another plane region (P for "pattern") with a distinguished point $D(P)$ called its focus. Let $X \sim Y$ mean that the plane sets X and Y are similar in the usual sense. Define now the sequence $C_0, C_1, C_2, \ldots$ of continuation regions by the recursion:

$$C_0 = I, \quad C_{n+1} = \{p : p \in Y, Y \sim P, D(Y) \in C_n, Y \cap F = \varnothing\}, \quad n = 0, 1, \ldots.$$

EXAMPLE 1. I is the origin o, F is a discrete set of points, P is an open circular disk whose focus is at the center. C_n is here the exact region to which an analytic function f is continuable by n successive power-series continuations, starting from I and supposing F to be the set of singularities.

EXAMPLE 2. I and F are as before, and P is the semi-open segment $[0, 1)$ with $(0, 0)$ as the focus. C_1 is now the principal star of f with respect to I [19].

EXAMPLE 3. I and F are arbitrary, P is of the form shown in Figure 1a with the focus p. P is supposed to represent an idealized antenna pattern showing the relative power radiated in various directions from p by a directional communication equipment at p. We suppose that the total power radi-

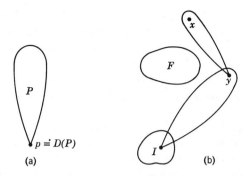

FIGURE 1.   Continuation and communication regions.

ated is arbitrarily variable and that changing it just scales P up or down. We suppose also that it is desired never to radiate into the forbidden region F for reasons of noninterference, secrecy, etc. (there are actually some further distance and threshold factors which we omit for simplicity). We cannot radiate from I to a point such as x (as shown in Figure 1b) directly but we can do it indirectly by radiating to y and placing at y a repeater (a copy of our directional communication equipment). Now C_n is the region into which we can radiate from I via a chain of n repeaters.

These examples show some generality of the formulation; the reader may wish to investigate the structure of the regions C_n and C_∞ with special reference to boundedness, shape, asymptotic shape, and smoothness of the boundaries.

As an example we indicate how to determine C_n for the case of analytic continuation of a function with one singularity. This corresponds to our Example 1 with F consisting of a single point—we take it to be $(1, 0)$. C_1 is then the open unit disk about the origin and C_2 is the union of all open disks with centers in C_1 and F on the boundary. It turns out that the boundary of C_2 is a cardioid; scaling it down radially with respect to F in the ratio $1 : 2$ gives us another cardioid which is the pedal curve of the boundary of C_1 with respect to F. The reader may wish to show that the boundary of C_{n+2} is a similarly scaled-up pedal curve of the boundary of C_{n+1} with respect to F. It follows by induction on n that the polar equation of the boundary of C_n, with F as the origin, is

$$r = 2^n \cos^{n+1}\left(\frac{\pi - \theta}{n + 1}\right), \qquad n \geq 2.$$

The reader may also wish to find the region C_n when the set F consists of several points on a circle about the origin; in this case the boundary of C_n consists of several types of arcs.

(c) A Prediction Problem. Here we use the continuation principle to solve the prediction problem for motion under curvature limitations. We suppose that a point p moves in the plane with constant speed v, starting at the time $t = 0$ from the origin tangentially to the positive y-axis. The motion of p is supposed to be a realistic model for some moving object. We suppose therefore that the path C traced by p satisfies the following conditions: (a) C is continuous (because p cannot jump through space), (b) C is continuously differentiable (because p has a velocity at each point, of magnitude v), (c) except for isolated points C has a continuous radius of curvature bounded from below by a positive constant r_0 (because p has a limited maneuverability and cannot turn too sharply). We now ask: What is the exact region $R(t)$ where p may land at time t?

Suppose for the time being that t satisfies $0 \le t \le \pi r_0/2v$. With reference to Figure 2 let C_1 and C_2 be the paths of tightest possible curvature; these are circles of radius r_0 tangent at the origin to the y-axis. Suppose that a thin

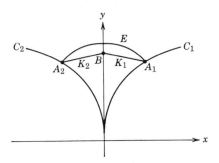

FIGURE 2.   Motion with curvature limitations.

inextensible string of length vt lies on C_1 reaching from the origin to the point A_1. When the string unwinds off C_1 onto C_2, remaining taut at all times, the moving end-point traces out an arc E with the end-points A_1 and A_2. E consists of two symmetric arcs of the evolute of the circle. It is clear that p can reach every point on E by an admissible path, i.e., by a path which satisfies the conditions (a), (b), and (c). Moreover, E is the front-part of the boundary of $R(t)$: it is the locus of all positions of p if p's object is to move away as far from the origin as is possible in a given direction. To find the back-part of the boundary of $R(t)$ we have to examine the possible positions of p if its object is to stay *close* to the origin. The reader may wish to show that now p starts moving on a circular arc of tightest possible curvature bending one way; this is followed by another such arc bending the other way. It follows that the back-part of the boundary of $R(t)$ consists of two symmetric arcs K_1 and K_2 of cardioids, shown in Figure 2.

Having determined $R(t)$ for small t we invoke the continuation principle to obtain $R(t)$ for all t. Let T satisfy $0 \leq T \leq \pi r_0/2v$; we use the continuation principle to determine $R(t + T)$ in the following way. Let C be an admissible path corresponding to the motion of p over the time interval from 0 to $t + T$. At the time t, p must be somewhere in the region $R(t)$ which we already know, say at the point x. Now, *starting from x and* in *the direction tangent to C at x*, we determine the region $R(T)$ which we also know. Then we take the union of all such regions $R(T)$ and this union is exactly $R(t + T)$. We save on the labor involved in finding $R(t + T)$ by observing what might be called the interiority principle of continuation: it is not necessary to take the union over all positions x but only over those points x which form the boundary $\partial R(t)$ of $R(t)$.

Examining our sketch of the derivation of $R(t)$ we find that with one exception every point of $\partial R(t)$ is accessible by just one admissible path. Thus, there is a unique arrival angle $u(x)$ which the path at x makes with the positive x-axis. The sole exception is the point B on the y-axis, shown in Figure 2, for which there are two symmetric admissible paths. We now relabel $R(t)$ as $R(t, o, \pi/2)$ to indicate the origin of the paths and the initial direction of motion. Making some provision for the ambiguity at B, we find now that

$$R(t + T, o, \pi/2) = \bigcup_{x \in \partial R(t, o, \pi/2)} R(T, x, u(x)).$$

The same procedure may be repeated and eventually we determine $R(t)$ for every t. It turns out, rather curiously, that taking $r_0 = v = 1$ for normalization purposes, the region $R(t)$ is simply connected if

$$0 \leq t < 1 + 3\pi/2 \qquad \text{or} \qquad 2\pi + \text{arc} \cos \frac{23}{27} \leq t$$

and doubly connected if

$$1 + 3\pi/2 \leq t < 2\pi + \text{arc} \cos \frac{23}{27}.$$

The graphs of $R(t)$ for several values of t are given in Figure 3 (where we assume that $r_0 = v = 1$); note that the scale varies from case to case in that figure. The reader who has some elements of physics may notice the formal resemblance of our continuation method as used in the above problem, and the Huygens principle.

Our prediction problem for motion under curvature limitations suggests the following pursuit problem. Two points P_1 and P_2 are moving in the plane subject to the following conditions: P_i moves with constant speed v_i along a smooth curve whose radius of curvature is $\geq r_i$, the motion starts with P_i at p_i moving in the initial direction u_i. Under what conditions will P_1 be

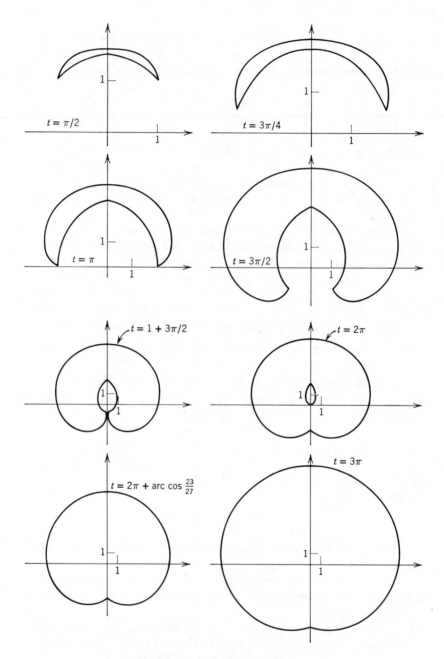

FIGURE 3. The regions $R(t)$.

able to catch P_2? The reader may wish to prove the following result due to Cockayne [13]: capture occurs for all initial configurations (p_1, p_2, u_1, u_2) if and only if $v_1 > v_2$ and $v_1{}^2 r_2 \geq v_2{}^2 r_1$.

We may also consider a pursuit problem in which P_2 moves with constant speed along a predetermined curve E while P_1 pursues P_2 by moving with constant speed and aiming at all times at the instantaneous position of P_2. The path P of P_1 is called the pursuit curve for the escape curve E. The reader may wish to prove the following smoothing property of the pursuit: if E is of the differentiability class C^k then P is of the differentiability class C^{k+1}. For further references and problems see [17].

Finally, there is a variety of pursuit problems in which we have a plane domain D, n pursuers $p_1, p_2, \ldots, p_n$, and one escapee p_0; all p's move in D along arbitrary differentiable curves, with constant speeds v. We must distinguish here between bounded and unbounded domains, and between capture and asymptotic capture.

7. ASYMPTOTIC ANALYSIS

This important branch of analysis deals with the limiting behavior of functions, integrals, sums of series, solutions of differential equations, etc., when certain variables or parameters tend to a limit or approach infinity. We have already handled some simple problems of this type:

(1) In Section 1 of Chapter 3, we have considered the problem of generating rationally the square root function, and we have found that if

$$x_1 = 1, \qquad x_{n+1} = x_n + 1/x_n, \qquad n = 1, 2, \ldots$$

then for large n

$$x_n = (2n)^{1/2} + o(1);$$

this suggests the possibility of an improved expansion such as, for instance,

$$x_n = (2n)^{1/2} + o(n^{-1} \log n) \qquad n \text{ large}$$

or

$$x_n = (2n)^{1/2} + cn^{-1} \log n + o(n^{-3/2} \log n) \qquad n \text{ large};$$

(2) In Section 6 of Chapter 2 we have estimated the nth iterate $f_n(x_0)$ of the sine function $f(x) = \sin x$; it was found that

$$f_n(x_0) = (3/n)^{1/2}[1 + o(n^{-1} \log n)].$$

The deeper analysis of de Bruin shows that

$$f_n(x_0) = (3/n)^{1/2} \left[1 - \frac{3 \log n}{10n} - \frac{c}{2n} + \right.$$
$$\left. n^{-2}(A \log^2 n + B \log n + C) + \frac{o(\log^3 n)}{n^3} \right]$$

where c is a constant depending on x_0 and

$$A = 27/200, \qquad B = 9c/20 - 9/50, \qquad C = 3c^2/8 - 3c/10 + 79/700.$$

An asymptotic power series for $f(x)$ has the form

$$f(x) \sim a_0 + a_1 x^{-1} + a_2 x^{-2} + \cdots,$$

the series may be divergent for all values of x but it is required that for every n

$$\lim_{x \to \infty} x^n \left| f(x) - \sum_{j=0}^{n} a_j x^{-j} \right| = 0.$$

More generally, a sequence of functions $f_1(x)$, $f_2(x)$, ... is asymptotic at $x = x_0$ if

$$\lim_{x \to x_0} f_{j+1}(x)/f_j(x) = 0 \qquad j = 1, 2, \ldots.$$

A function $f(x)$ has an asymptotic representation

$$f(x) \sim \sum_{n=1}^{\infty} a_n f_n(x)$$

with respect to this sequence if

$$\lim_{x \to x_0} \left| f(x) - \sum_{j=1}^{n} a_j f_j(x) \right| \bigg/ f_n(x) = 0, \qquad n = 1, 2, \ldots.$$

To see how an asymptotic series, though possibly divergent, may be used to calculate the value of the function it represents, we consider the integral

$$E(x) = \int_x^\infty t^{-1} e^{-t} \, dt$$

which cannot be evaluated in terms of elementary functions. Integrating by parts n times we have

$$E(x) = e^{-x}[x^{-1} - 1! \, x^{-2} + 2! \, x^{-3} - \cdots + (-1)^{n-1}(n-1)! \, x^{-n}]$$
$$+ (-1)^n n! \int_x^\infty t^{-n-1} e^{-t} \, dt \quad (1)$$

which gives us the asymptotic expansion (near $x = \infty$)

$$E(x) \sim e^{-x} \sum_{n=0}^{\infty} (-1)^n n! \, x^{-n-1}. \tag{2}$$

We observe that the series above diverges for all x. The remainder term in (1) is estimated as follows:

$$\left| n! \int_x^\infty t^{-n-1} e^{-t} \, dt \right| < n! \int_x^\infty x^{-n-1} e^{-t} \, dt = n! \, x^{-n-1} e^{-x}$$

so that

$$\left| E(x) - e^{-x} \sum_{j=0}^{n-1} (-1)^j j! \, x^{-j-1} \right| < n! \, x^{-n-1} e^{-x}. \tag{3}$$

Therefore, for a fixed n there is $x_0 = x_0(n)$ such that the nth partial sum represents $E(x)$ with arbitrarily small truncation error for $x \geq x_0$. Conversely, for a fixed x there is a value $n_0 = n_0(x)$ which gives us a stopping rule (tells us not to add any more terms); here $n_0 = [x]$ or $[x] - 1$. For, the terms of the sum in (2) decrease to a minimum and then start increasing again. In our example we have the special circumstance of alternating series behavior: by (1) and (3) the truncation error is less than the absolute value of the first term left out.

Another example of an asymptotic series was the Stirling formula of Section 13, Chapter 3:

$$\log \frac{n!}{\sqrt{2\pi n}(n/e)^n} \sim \sum_{k=1}^\infty \frac{(-1)^{k-1} B_{2k+1}}{(2k+1)(2k+2)} n^{-2k-1}$$

where B_j is the jth Bernoulli number. By taking exponentials it was found that

$$n! \sim \sqrt{2\pi n}\left(\frac{n}{e}\right)^n \left[1 + \frac{1}{12n} + \frac{1}{288n^2} - \frac{139}{51840n^3} + \cdots\right]. \tag{4}$$

As an example of the Laplace method of asymptotic expansion of integrals, we derive a series which generalizes (4). We start with the Euler integral

$$\Gamma(x+1) = \int_0^\infty u^x e^{-u} \, du = \int_0^\infty e^{-u + x \log u} \, du$$

and we reason as follows. The exponent $-u + x \log u$ attains its maximum at $u = x$; supposing that x is large the principal contribution to the value of the integral comes from those u which are close to x. The integrand already has a sharp peak at $u = x$; to sharpen that peak further we apply a preliminary transformation $u = xt$ giving us

$$\Gamma(x+1) = x^{x+1} \int_0^\infty e^{x(-t + \log t)} \, dt. \tag{5}$$

Further substitution is made:

$$t - \log t = v^2 + 1; \tag{6}$$

either by the Bürmann–Lagrange–Taylor formula or by the longer but more straightforward method of undetermined coefficients we find that

$$t = 1 + 2^{1/2}v + 2/3v^2 + 2^{-1/2}/9v^3 - 2/135v^4 + 2^{-1/2}/540v^5 - \cdots.$$

When this and (6) are used in (5) we find, noting that the odd powers may be dropped,

$$\Gamma(x + 1) = x^{x+1}e^{-x} \int_{-\infty}^{\infty} e^{-xv^2}\left[\sqrt{2} + \frac{1}{3\sqrt{2}}v^2 + \frac{1}{108\sqrt{2}}v^4 + \cdots\right] dv.$$

We have

$$\int_{-\infty}^{\infty} e^{-xv^2}\,dv = \sqrt{\pi}x^{-1/2}$$

and by differentiating with respect to the parameter x,

$$\int_{-\infty}^{\infty} v^2 e^{-xv^2}\,dv = \frac{1}{2}\sqrt{\pi}x^{-3/2}, \qquad \int_{-\infty}^{\infty} v^4 e^{-xv^2}\,dv = \frac{3}{4}\sqrt{\pi}x^{-5/2}.$$

Using these, we find the asymptotic series

$$\Gamma(x + 1) \sim \sqrt{2\pi x}(x/e)^x\left[1 + \frac{1}{12}x^{-1} + \frac{1}{288}x^{-2} + \cdots\right]. \tag{7}$$

The following theorem on divergent series is due to Cesaro [38]: let $\sum_{n=0}^{\infty} b_n$ be a divergent series of nonnegative terms, let $a_n \geq 0$ for all n and let

$$A_n = \sum_{j=0}^{n} a_j, \qquad B_n = \sum_{j=0}^{n} b_j;$$

suppose that either $\lim_{n=0} a_n/b_n = L$ or $\lim_{n=0} A_n/B_n = L$; then

$$\lim_{x \to 1-0}\left(\sum_0^{\infty} a_n x^n\right)\bigg/\left(\sum_0^{\infty} b_n x^n\right) = L.$$

We consider two applications to asymptotic analysis.

EXAMPLE 1. Estimate $\sum_{n=0}^{\infty} x^{n^2}$ for the values of x close to 1. We apply Cesaro's theorem with $a_n = 0$ if n is not a perfect square and $a_n = 1$ otherwise, so that

$$A_n = [\sqrt{n}] + 0(1).$$

For $\sum_{n=0}^{\infty} b_n x^n$ we take a possibly simple series whose nth partial sum is of the order $\sqrt{n}$ at $x = 1$:

$$\sum_{n=0}^{\infty} b_n x^n = (1 - x)^{-1/2}.$$

B_n is then the coefficient of x^n in the series for $(1 - x)^{-1} (1 - x)^{-1/2}$ so that

$$B_n = (-1)^n \binom{-3/2}{n} = \frac{3 \cdot 5 \cdot 7 \cdots (2n + 1)}{2 \cdot 4 \cdot 5 \cdots (2n)}.$$

Using the Wallis product

$$\frac{\pi}{2} = \prod_{n=1}^{\infty} \frac{(2n)(2n)}{(2n - 1)(2n + 1)}$$

we have

$$\lim_{n \to \infty} A_n/B_n = \sqrt{\pi/2}$$

and therefore by Cesaro's theorem, as $x \to 1$

$$\sum_{n=0}^{\infty} x^{n^2} \sim \frac{\sqrt{\pi}}{2} (1 - x)^{-1/2}.$$

EXAMPLE 2. Let $a > 1$ be an integer, estimate $\sum_{n=0}^{\infty} x^{a^n}$ for x close to 1. We proceed as before with

$$A_n \sim \log n/\log a,$$

and we take

$$\sum_{n=1}^{\infty} b_n x^n = \log(1 - x)$$

so that $B_n \sim -\log n$. By Cesaro's theorem we find the asymptotic estimate near $x = 1$

$$\sum_{n=0}^{\infty} x^{a^n} \sim -\frac{1}{\log a} \log(1 - x).$$

We shall obtain next Hardy's estimate of the size of the partition function $p(n)$ [27]. As in Section 2, we have

$$1 + \sum_{n=1}^{\infty} p(n)x^n = \prod_{n=1}^{\infty} (1 - x^n)^{-1} = F(x)$$

say. Taking logarithms, using Lambert series, and interchanging the order of summations, we have

$$\log F(x) = \sum_{n=1}^{\infty} \log(1 - x^n)^{-1} = \sum_{n=1}^{\infty} \sum_{m=1}^{\infty} x^{mn}/m = \sum_{m=1}^{\infty} \frac{1}{m} \frac{x^m}{1 - x^m}.$$

By the mean value theorem

$$m x^{m-1}(1 - x) < 1 - x^m < m(1 - x), \qquad 0 < x < 1$$

so that

$$\frac{1}{m} < \frac{1-x}{1-x^m} < \frac{1}{m} x^{1-m}$$

and hence

$$\sum_{m=1}^{\infty} x^m/m^2 < (1-x)\log F(x) < x \sum_{m=1}^{\infty} m^{-2} = \pi^2 x/6.$$

Therefore for x close to 1

$$F(x) = e^{[o(1)+\pi^2/6]/(1-x)}. \tag{8}$$

Now we look for the approximate order of magnitude of $p(n)$. It turns out that it grows faster than a power of n for if $p(n) \sim Kn^a$ for large n then an easy application of Cesaro's theorem would give us

$$F(x) = 1 + \sum_{n=1}^{\infty} p(n)x^n \sim K\Gamma(1+a)(1-x)^{-a-1}$$

near $x = 1$, contradicting (8). On the other hand, $p(n)$ cannot grow as fast as c^n for any $c > 1$, since then $F(x)$ would not be convergent, for $|x| < 1$. Following Hardy, we suppose that

$$p(n) \sim e^{Bn^b}, \qquad B > 0, \qquad 0 < b < 1, \qquad n \text{ large},$$

and we reason as follows. First,

$$F(x) = 1 + \sum_{n=1}^{\infty} e^{Bn^b} x^n = 1 + \sum_{n=1}^{\infty} e^{Bn^b - ny} \tag{9}$$

where $x = e^{-y}$. The largest term of this series is the one for which

$$Bbn^{b-1} = y$$

approximately, and so this largest term is very nearly

$$e^{C(1-x)^{-b(1-b)-1}}$$

where

$$C = B^{(1-b)^{-1}} b^{b(1-b)^{-1}} (1-b).$$

Next, the whole sum $F(x)$ in (9) is of the same order of magnitude as the largest term alone, the difference being chargeable to the $o(1)$ term of (8); equating the above and (8) we find that

$$b = 1/2, \qquad B = \pi(2/3)^{1/2}$$

so that, for large n,

$$p(n) \sim e^{\pi(2n/3)^{1/2}}.$$

Actually, the above analysis is too crude to detect possible factors whose size is much smaller than the exponential order; the correct asymptotic expression for $p(n)$ is

$$p(n) \sim (4n \cdot 3^{1/2})^{-1} e^{\pi(2n/3)^{1/2}}.$$

In the remainder of this section we deal with a topic loosely related to asymptotic analysis; our main purpose is to make a conjecture and to suggest a number of problems. It was shown before that the square root function is rationally generable in a certain asymptotic sense. We ask now: What other functions are similarly generable? To be more precise we shall connect our development to that of Section 2 in Chapter 4; we introduce now some terminology. Let $p_i(n)$, $i = 1, 2, \ldots, k$, $n = 0, 1, \ldots$ be k functions given by the polynomial recursive scheme

$$p_i(n + 1) = P_i[p_1(n), \ldots, p_k(n)], \qquad n = 0, 1, \ldots, p_i(0) = a_i \qquad (10)$$

where a_i are given integers and P_i are polynomials with integer coefficients, in k variables. We say that each such $p_i(n)$ is a PRF (polynomially recursive function) of order $\leq k$, it is a PRF of order exactly k if no smaller integer will do for k. It is easy to show that the following functions are PRF's: $n!$, 3^n, 2^{3^n},

$$f(n) = \prod_{k=1}^{n} [P(k!)]^{n+1-k}, \qquad (11)$$

where P is a fixed polynomial with integer coefficients. On the other hand, it is likely that the following functions are not PRF's:

$$n^n, (n^2)!, [n^{1/2}], n\text{th consecutive prime}.$$

The reason for this difference in behavior is (likely to be) as follows. The factorial $n!$ is a PRF because one-past-value memory reduces it to something simpler: $(n + 1)! = (n + 1)n!$. Similarly, with the more complicated example, (11), we have $f(n + 1) = f(n)g(n)$ where

$$g(n) = \prod_{k=1}^{n+1} P(k!);$$

and continuing with the simplification process,

$$g(n + 1) = g(n)P[(n + 2)!].$$

Thus, having two-past-values memory reduces the problem of generating $f(n)$ to that of generating $P(n!)$. Since $n!$ is a PRF and the sums and products of PRF's are themselves PRF's, it follows that $P(n!)$ and so $g(n)$ and $f(n)$ are also PRF's.

On the other hand, no finite amount of past-values memory appears to reduce n^n or $(n^2)!$ to anything simpler than themselves. There is a trivial order-of-growth condition which excludes certain functions from being PRF's: any solution $p_i(n)$ of the system (10) grows no faster than a^{b^n} where a and b are some positive constants. But no other general criterion appears to be known. It is remotely possible that there is some connection here with the phenomenon of functional incompatibility: one function cannot satisfy two functional equations of "different" types. The only exception appears to be the class of exponential polynomials, defined in Section 5; each such exponential polynomial satisfies a linear differential equation with constant coefficients as well as a linear difference equation with constant coefficients. A prime example of functional incompatibility is the theorem of Holder stated at the end of Chapter 5: the Γ-function, in virtue of satisfying the difference equation

$$\Gamma(x + 1) = x\Gamma(x),$$

does not satisfy any algebraic differential equation.

To be able to handle in the same manner functions such as $n^{1/2}$ which are not integer-valued, we extend now the definition of a PRF. A function $\phi(n)$ is said to be APRF (asymptotically polynomially recursive function) if there exists a system of the type (10) such that

$$p_1(n)/p_2(n) - \phi(n) \to 0 \qquad \text{as } n \to \infty.$$

The order of an APRF is defined in the same way as for a PRF. What we have previously, in Section 2 of Chapter 4, called an AM-number becomes now simply an APRF which is a constant. It was shown that if

$$x_1 = 1, \qquad x_{n+1} = x_n + 1/x_n, \qquad n = 1, 2, \ldots$$

then for large n

$$x_n - (2n)^{1/2} \to 0.$$

Using the APR constant $2^{1/2}$, it is not hard to show that $n^{1/2}$ is an APRF. An alternative proof, which the reader may wish to supply, could proceed by the asymptotic analysis of the variable form of the Newton–Raphson algorithm for the square root: the reader may wish to show that if

$$x_1 = 1, \qquad x_{n+1} = \frac{1}{2}(x_n + n/x_n) \qquad n = 1, 2, \ldots$$

then

$$x_n - n^{1/2} \to 0 \qquad \text{as } n \to \infty.$$

Once this is done, one lets $x_n = p(n)/q(n)$ and the relations

$$p(n + 1) = p^2(n) + nq^2(n), \qquad q(n + 1) = 2p(n)q(n)$$

lead to the defining system of order 3 for $n^{1/2}$:

$$p(n + 1) = p^2(n) + q^2(n)r(n), \qquad q(n + 1) = 2p(n)q(n),$$
$$r(n + 1) = r(n) + 1, \qquad p(0) = q(0) = 1, \qquad r(0) = 0.$$

Similarly, any fractional power $n^{s/t}$ (with s and t positive coprime integers) is APRF, as can be shown by a suitable encoding of the Newton–Raphson algorithm for the tth root. Somewhat more generally, let $\phi(n)$ be a well-defined real-valued function, algebraic over the integers. That is, $y = \phi(n)$ satisfies an equation of the form

$$P_0(n)y^k + P_1(n)y^{k-1} + \cdots + P_{k-1}(n)y + P_k(n) = 0$$

where each P_i is a polynomial with integer coefficients. Then, using for instance the theorem that an algebraic function has a fractional power series, the reader may wish to show that $\phi(n)$ is an APRF.

It would be of considerable interest to find out which transcendental functions are APRF's and which are not. We conjecture, for instance, that the function $\log n$ is not an APRF, and we note that the truth of our conjecture implies not only that the Euler constant γ is transcendental, but further that it is algebraically independent of 1, π, e, $\log 2$, $\zeta(3)$, and $J_0(1)$ simultaneously. This implication is not very hard to show since

(a) $\gamma = \lim\limits_{n \to \infty} \left(\sum\limits_{j=1}^{n} 1/j - \log n \right)$,

(b) $\sum\limits_{j=1}^{n} 1/j$ is an APRF (in fact, a quotient of two PRF's),

(c) the transcendental constants above, as well as $\zeta(3)$, are APRF's being (in the previous terminology) AM-numbers, and

(d) the proof, given in Section 2 of Chapter 4, that a real irrational number is algebraic if and only if it is of order 2 extends without undue difficulty to a proof of the following: if x and y are real irrational numbers of orders p and q where $2 \le p < q - 2$, then x and y are algebraically independent.

8. COINCIDENCES, FORBIDDEN CONFIGURATIONS, AND HYPERGRAPHS

(a) We have met graphs several times already. A graph G consists of n points $v_1, \ldots, v_n$ called vertices, and a number of edges $v_i v_j$, subject to the conditions that no edge joins a vertex to itself and no two vertices are joined by more than one edge. Occasionally these two conditions are not imposed, as in Section 6 of Chapter 4; we speak then of a generalized graph. Otherwise, G is a geometrical picture of a binary irreflexive symmetric relation R on the set of the n vertices. R irreflexive means that no x stands in the relation R to itself; R symmetric means that if x stands in the relation R to y then y stands in the relation R to x. If the symmetry is dropped we must distinguish between $v_i v_j$ and $v_j v_i$, and we get then a directed graph. G is complete if every two vertices are connected by an edge; it is connected if every two vertices are connected by a sequence of edges; and it is a tree if every two vertices are connected by a unique sequence of edges. It follows that a tree is a connected graph without closed circuits.

A complete graph on n vertices $v_1, \ldots, v_n$ may be thought of as the set of all $\binom{n}{2}$ edges, or one-dimensional faces, of an $(n-1)$-dimensional simplex. Since every graph G is obviously a subgraph of the complete graph on the same vertices, every G may be regarded as a subset of the $\binom{n}{2}$ one-dimensional faces of an $(n-1)$-dimensional simplex. Renaming graphs as 1-graphs, we are led to define a k-dimensional hypergraph, or briefly a k-graph, on n vertices $v_1, \ldots, v_n$; $0 \le k \le n-1$. Such a k-graph is just a subset of the $\binom{n}{k+1}$ k-dimensional faces of the simplex with the vertices $v_1, \ldots, v_n$. We may also view a k-graph as a geometrical picture of a $(k+1)$-ary relation R which is totally irreflexive and totally symmetric (totally irreflexive means: if x_1, $x_2, \ldots, x_{k+1}$ are related by R then they must be all distinct; totally symmetric means: if $x_1, x_2, \ldots, x_{k+1}$ are related by R then so is any permutation of the x's).

In this section we show how k-graphs arise naturally in a general problem concerning combinatorial probabilities and forbidden configurations. Since the problem is rather complicated we consider first a special case which is of interest in its own right.

(b) On the interval $[0, L]$ n points $x_1, \ldots, x_n$ are taken independently and uniformly at random; what is the probability $P = P(n, a, L)$ that no two points are closer than a? We assume that $(n-1)\,a < L$ since otherwise $P = 0$. The n points can be ordered in $n!$ ways, suppose that we have $x_1 \le x_2 \le \cdots \le x_n$. Then the conditions of the problem are satisfied if and only if

$$0 \le x_1 \le x_2 - a \le x_3 - 2a \le \cdots \le x_n - (n-1)a \le L - (n-1)a. \quad (1)$$

Let $y_i = x_i - (i-1)a$, then the probability that (1) holds is L^{-n} times the volume of the region of n-tuples $y = (y_1, \ldots, y_n)$ for which

$$0 \le y_1 \le y_2 \le \cdots \le y_n \le L - (n-1)a.$$

This volume is

$$\int_0^{L-(n-1)a} \int_0^{y_n} \cdots \int_0^{y_3} \int_0^{y_2} dy_1\, dy_2 \cdots dy_{n-1}\, dy_n = \frac{1}{n!}[L - (n-1)a]^n.$$

Multiplying this by the number $n!$ of equiprobable orderings and dividing by L^n we get

$$P = [1 - (n-1)a/L]^n. \quad (2)$$

The probability P could also be found as follows. Let $n = 2$ first, the sample space of pairs $x = (x_1, x_2)$ is the Cartesian product, i.e., the square $S = [0, L] \times [0, L]$, the diagonal D is given by $x_1 = x_2$, and the hexagonal set H_{12} shown in Figure 1a is the locus of all $x = (x_1, x_2)$ such that $|x_1 - x_2| \leq a$.

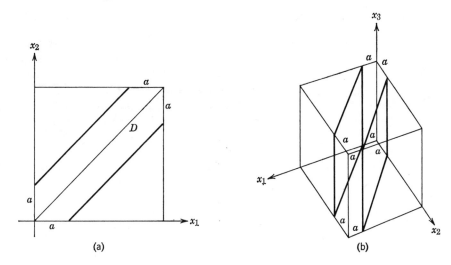

FIGURE 1.   Coincidence in a square and a cube.

When H_{12} is removed from S the remaining two triangles fit together to make up a smaller square of side $L - a$. Since x_1 and x_2 are independent and uniformly at random on $[0, L]$, it follows that $x = (x_1, x_2)$ is uniformly at random in the square S. Therefore, P is just the ratio of the areas of the two squares: $P = (L - a)^2/L^2$, which is (2) with $n = 2$. Similarly, for $n = 3$ we consider the cube H of Figure 1b and in its square face S_{12} we form H_{12} as before. Let B_{12} be the cylinder of height L on H_{12} as base, shown in Figure 1b. It is the locus of triples (x_1, x_2, x_3) which are characterized by the too near approach of x_1 and x_2. B_{13} and B_{23} are similarly constructed and now the sample space of allowed triples (x_1, x_2, x_3) is

$$H - (B_{12} \cup B_{13} \cup B_{23}).$$

In the case of general n we let H be the n-dimensional cube

$$H = \{(x_1, \ldots, x_n) : 0 \leq x_i \leq L, i = 1, \ldots, n\};$$

S_{ij} is the two-dimensional face of H given by

$$S_{ij} = \{(x_1, \ldots, x_n) : x_k = 0 \text{ if } k \neq i, k \neq j, 0 \leq x_i \leq L. 0 \leq x_j \leq L\};$$

H_{ij} is the hexagonal subset of S_{ij} and B_{ij} is the cylinder on H_{ij} as base:

$$B_{ij} = \{(x_1, \ldots, x_n) : 0 \le x_k \le L, 1 \le k \le n, |x_i - x_j| \le a\}.$$

Now the sample space of allowed configurations is

$$H - \bigcup_{1 \le i < j \le n} B_{ij}. \tag{3}$$

The reader may wish to show that when the $\binom{n}{2}$ sets B_{ij} are removed from H there remain $n!$ congruent simplexes which can be moved together so as to form a smaller n-dimensional cube of side-length $L - (n - 1)a$. As before, P is the ratio of the volumes of the two cubes and we get (2) again.

The reader may wish to consider the following generalization. Let $A = (a_{ij})$ be an n by n matrix of real numbers which satisfy the following conditions:

$$a_{ii} = 0 \qquad i = 1, \ldots, n,$$
$$a_{ij} \ge 0 \qquad i, j = 1, \ldots, n,$$
$$a_{ij} + a_{jk} \ge a_{ik} \qquad i, j, k = 1, \ldots, n.$$

Let S be the symmetric group on n elements with members written as permutations

$$\sigma = \begin{pmatrix} 1 & 2 & \cdots & n \\ \sigma(1) & \sigma(2) & \cdots & \sigma(n) \end{pmatrix}.$$

If n labeled points $x_1, x_2, \ldots, x_n$ are taken independently and uniformly at random on $[0, L]$ then the probability that $|x_i - x_j| \ge a_{ij}$ for all i and j is

$$\frac{1}{n!} \sum_{\sigma \in S} \left[\max\left(0, 1 - L^{-1} \sum_{k=1}^{n-1} a_{\sigma(k)\sigma(k+1)}\right) \right]^n.$$

(c) Our general problem is as follows. The interval $[0, L]$ is replaced by a set A in the m-dimensional Euclidean space, $B(x)$ is the solid m-dimensional ball of diameter a centered at x, n points $x_1, \ldots, x_n$ are chosen in A independently and with the same probability density function f, defined on A. We consider the n balls $B(x_1), \ldots, B(x_n)$ and we ask for the probability $P = P(A, a, f, n, k)$ that no point in A belongs to k or more of the n balls. Actually, a much more general formulation is possible [48] but the above form will be sufficient for us. In the special case when $m = 1$, $A = [0, T]$, and $f = f(t)$ is a probability density on $[0, T]$, we have n events which occur independently during the time interval 0 to T with the density $f(t)$. If a k-fold coincidence means that k or more events occur during a time interval of length a, then P is the probability of no such k-fold coincidence. This shows that the general problem has some relevance to telephone and vehicle traffic, serving and queuing lines, multiple particle counters, and in similar contexts.

We consider now another special case, related to the problem of deriving a van der Waals type of an equation of state from a primitive hard-sphere gas model. Here $m = 3$, A is the three-dimensional cube of side L, $n = 2$, $k = 2$, and f is uniform over A. P is now the probability that two points independently and uniformly at random in A are no closer than a. The reader may wish to show that

$$1 - P = L^{-6} \int_{m_6}^{M_6} \int_{m_5}^{M_5} \cdots \int_{m_1}^{M_1} dx_1 \, dx_2 \, dy_1 \, dy_2 \, dz_1 \, dz_2 \tag{4}$$

where

$$m_1 = \max\{0, x_2 - [a^2 - (y_1 - y_2)^2 - (z_1 - z_2)^2]^{1/2}\},$$
$$M_1 = \min\{L, x_2 + [a^2 - (y_1 - y_2)^2 - (z_1 - z_2)^2]^{1/2}\},$$
$$m_3 = \max\{0, y_2 - [a^2 - (z_1 - z_2)^2]^{1/2}\},$$
$$M_3 = \min\{L, y_2 + [a^2 - (z_1 - z_2)^2]^{1/2}\},$$
$$m_5 = \max\{0, z_2 - a\},$$
$$M_5 = \min\{L, z_2 + a\},$$
$$m_2 = m_4 = m_6 = 0, \qquad M_2 = M_4 = M_6 = L.$$

The arrangement of limits of integration in (4) corresponds to two points (x_1, y_1, z_1) and (x_2, y_2, z_2) as centers of spheres of radius $a/2$; one point moves freely over the cube while the other one moves over the cube so that the spheres intersect. If $a \le L$ the sixfold integral in (4) is evaluable in terms of elementary functions and the reader with good staying power may wish to verify that then

$$P = 1 - \frac{4\pi}{3}(a/L)^3 + \frac{3\pi}{2}(a/L)^4 - \frac{8}{5}(a/L)^5 + \frac{1}{6}(a/L)^6.$$

(d) For the general problem we shall obtain the answer in the form analogous to (3) except that the equivalents of H and of the sets B_{ij} will serve as regions of integrations for f, and instead of the B_{ij}'s we shall have similar quantities with k rather than two indices. Our procedure is best explained by evaluating the probability P of (2) for the third time. We observed before that P is the volume of the set given by (3), divided by L^n:

$$P = 1 - L^{-n} \operatorname{Vol}\left(\bigcup_{1 \le i < j \le n} B_{ij}\right). \tag{5}$$

We now find P by evaluating the volume of the union

$$V = \bigcup_{1 \le i < j \le n} B_{ij} \tag{6}$$

by means of the inclusion-exclusion principle. This principle states that if $X_1, X_2, \ldots, X_N$ is a finite collection of sets then

$$
m\left(\bigcup_1^N X_i\right) = \sum_{1 \le i_1 \le N} m(X_{i_1}) - \sum_{1 \le i_1 < i_2 \le N} m(X_{i_1} \cap X_{i_2})
$$
$$
+ \sum_{1 \le i_1 < i_2 < i_3 \le N} m(X_{i_1} \cap X_{i_2} \cap X_{i_3})
$$
$$
- \cdots + (-1)^{N+1} \sum_{1 \le i_1 < \cdots < i_N \le N} m(X_{i_1} \cap \cdots \cap X_{i_N}). \quad (6a)
$$

Here $m(X)$ is a measure of X and the two most often used cases are: (a) all sets are finite and $m(X)$ is the number of elements of X, and (b) all sets are sufficiently regular and $m(X)$ is the volume of X. Generalizing the second case somewhat, we may suppose that an integrable function f is defined on the union of the sets and $m(X)$ is the integral of f over X.

The $\binom{n}{2}$ sets B_{ij} can be re-enumerated with a single index as B_k where $k = 1, 2, \ldots, N$, with $N = \binom{n}{2}$. Applying the inclusion-exclusion principle to the sets B_k we find

$$
\text{Vol}(V) = \sum_{1 \le k_1 \le N} \text{Vol}(B_{k_1}) - \sum_{1 \le k_1 < k_2 \le N} \text{Vol}(B_{k_1} \cap B_{k_2}) + \cdots
$$

which we write in a self-explanatory terminology as

$$
\text{Vol}(V) = K_1 - K_2 + K_3 - \cdots + (-1)^{N+1} K_N. \quad (7)
$$

The first term is

$$
K_1 = \sum_{k=1}^{N} \text{Vol}(B_k)
$$

and since all sets B_k are congruent we have

$$
K_1 = N \, \text{Vol}(B_1) = N \, L^{n-2} \, \text{Area}(H_{12}). \quad (8)
$$

The second term in (7) is

$$
K_2 = \sum_{1 \le k_1 < k_2 \le N} \text{Vol}(B_{k_1} \cap B_{k_2});
$$

recalling the congruence of the sets B_k and their being really two-index quantities, we note that there is no need to sum in the above expression over all the

$$
\binom{N}{2} = \left(\!\!\binom{\binom{n}{2}}{2}\!\!\right)
$$

pairs of pairs of indices. For, there are only two *types* of B-set intersections taken two at a time: the type $B_{12} \cap B_{34}$ with no indices shared and the type $B_{12} \cap B_{13}$ with one index shared. There are

$$N_{21} = n(n-1)(n-2)(n-3)/8$$

intersections of the first type and

$$N_{22} = n(n-1)(n-2)/2$$

intersections of the second type. As a check we find

$$N_{21} + N_{22} = \left(\binom{n}{2} \atop 2 \right).$$

Accordingly,

$$K_2 = N_{21} \, \text{Vol}(B_{12} \cap B_{34}) + N_{22} \, \text{Vol}(B_{12} \cap B_{13}).$$

Since B_{ij} is a cylinder based on the hexagonal set H_{ij} we have

$$\text{Vol}(B_{12} \cap B_{34}) = L^{n-4} \, \text{Area}^2(H_{12}), \quad \text{Vol}(B_{12} \cap B_{13}) = L^{n-3} \, \text{Vol}(H_{12} \cap H_{13}).$$

Hence

$$K_2 = N_{21} L^{n-4} \, \text{Area}^2(H_{12}) + N_{22} L^{n-3} \, \text{Vol}(H_{12} \cap H_{13}). \tag{9}$$

Suppose that we wish to compute similarly the next term K_3. Already the matters grow somewhat complicated, and we need some machinery for distinguishing the various types of index-sharing for triples of pairs (i, j). There are five such types here; these are shown, together with their quite simply and naturally defined *incidence graphs*, in Figure 2. We let N_{3i} be the number of triples of pairs of the type i, $i = 1, \ldots, 5$. Then

$$K_3 = N_{31} L^{n-6} \, \text{Area}^3(H_{12}) + N_{32} L^{n-5} \, \text{Area}(H_{12})\text{Vol}(H_{12} \cap H_{13})$$

$$+ N_{33} L^{n-4} \, \text{Vol}(H_{12} \cap H_{23} \cap H_{34}) + N_{34} L^{n-4} \, \text{Vol}(H_{12} \cap H_{13} \cap H_{14})$$

$$+ N_{35} L^{n-3} \, \text{Vol}(H_{12} \cap H_{13} \cap H_{23}). \tag{10}$$

We shall now prepare some machinery which will allow us to handle similarly any term K_r. Let the five incidence graphs of Figure 2 be denoted by G_{3i}, $i = 1, \ldots, 5$. If G is an incidence graph let $v(G)$ be the number of its vertices, and let these vertices be enumerated in some order as $1, 2, \ldots, v(G)$. By the cluster integral $I(G)$ corresponding to the incidence graph G we understand the $v(G)$-fold integral

$$\int \cdots \int dx_1 \, dx_2 \cdots dx_{v(G)}, \tag{11}$$

taken over the domain which satisfies the following conditions:

(a) $0 \le x_i \le L$, $i = 1, \ldots, v(G)$,

(b) $|x_i - x_j| \le a$ if the vertex i and the vertex j are connected by an edge in G.

Now (10) may be written as

$$K_3 = \sum_{i=1}^{5} N_{3i} L^{n - v(G_{3i})} I(G_{3i}).$$

One further reduction will be applied: due to the simplicity of the integrand in (11) we notice that when an incidence graph G has components G_1 and G_2,

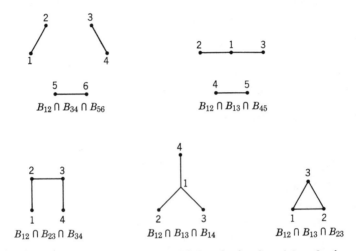

FIGURE 2.   Incidence graphs and index-sharing for triples of pairs.

as for instance, the second graph in Figure 2, then we have the multiplicative property

$$I(G) = I(G_1)I(G)_2 \tag{12}$$

for the corresponding cluster integrals. Therefore generally

$$I(G) = \prod_{C \subseteq G} I(C)$$

where C ranges over the connected components of G. Hence we have the final form for K_3:

$$K_3 = \sum_{i=1}^{5} N_{3i} L^{n - v(G_{3i})} \prod_{C \subseteq G_{3i}} I(C).$$

It is now relatively simple to obtain the expression for the general term K_r in (7). Let $t(r)$ be the number of different types of r-tuples of pairs, represented by their incidence graphs G_{ri}, $i = 1, \ldots, t(r)$. As before, let $v(G_{ri})$ be the number of vertices of G_{ri}. Let $N_{ri}(n)$ be the number of distinct r-tuples of pairs of the type i. Let the cluster integrals $I(C)$, with C ranging over connected components of G_{ri} be defined as before. Then

$$K_r = \sum_{i=1}^{t(r)} N_{ri}(n) L^{n - v(G_{ri})} \prod_{C \subseteq G_{ri}} I(C). \tag{13}$$

From (5), (6), (7), and (13) we get finally

$$P = 1 - \sum_{r=1}^{\binom{n}{2}} \sum_{i=1}^{t(r)} [(-1)^{r+1} N_{ri}(n) L^{-v(G_{ri})} \prod_{C \subseteq G_{ri}} I(C)]. \tag{14}$$

(e) We have now calculated the probability P of equation (2) for the third time and in a way which may appear needlessly laborious. But, as the reader will have guessed, this third method extends and generalizes so far that it will enable us to solve our general problem, that of finding the probability $P = P(A, a, f, n, k)$. Moreover, P will be found in a form completely analogous to (14).

We start with the analogues of the sets H, S_{ij}, B_{ij}, H_{ij}. Our interest is in the n-tuple $(x_1, \ldots, x_n)$ where each x_i ranges over A, thus instead of the n-dimensional cube H we have now the nm-dimensional Cartesian product

$$A^n = \{(x_1, \ldots, x_n) : x_i \in A, i = 1, \ldots, n\}.$$

If $1 \leq i_1 < i_2 < \cdots < i_k \leq n$ then the square face S_{ij} of H is replaced by the set

$$S_{i_1 \cdots i_k} = \{(x_1, \ldots, x_n) : x_{i_s} = 0 \text{ if } s \neq 1, \ldots, k; \, x_j \in A \text{ otherwise}\}.$$

Instead of the hexagonal set H_{ij} we have now the set

$$H_{i_1 \cdots i_k} = \{(x_1, \ldots, x_n) : |x_{i_s} - x_{i_p}| \leq a \text{ for } 1 \leq s < p \leq k, \, x_j = 0 \text{ otherwise}\}$$

and instead of the cylinder B_{ij} we have the cylinder

$$B_{i_1 \cdots i_k} = \{(x_1, \ldots, x_n) : x_j \in A, j = 1, \ldots, n, \, |x_{i_s} - x_{i_p}| \leq a, 1 \leq s < p \leq k\}.$$

Instead of (5) we have now

$$P = P(A, a, f, n, k) = 1 - \int \cdots \int_{\substack{\bigcup B_{i_1 \cdots i_k} \\ 1 \leq i_1 < \cdots < i_k \leq n}} \prod_{j=1}^{n} f(x_j) \, dx_1 \cdots dx_n. \tag{15}$$

To process this further we re-number the $N = \binom{n}{k}$ multi-indexed sets $B_{i_1 \cdots i_k}$ as B_j where $j = 1, \ldots, N$. We next use the inclusion-exclusion principle, to obtain

$$P = 1 - K_1 + K_2 - \cdots + (-1)^N K_N, \tag{16}$$

$$K_r = \sum_{1 \leq j_1 < \cdots < j_r \leq N} \cdots \sum \int \cdots \int_{B_{r1} \cap \cdots \cap B_{rr}} \prod_{j=1}^{n} f(x_j) \, dx_1 \cdots dx_n. \tag{17}$$

As before, we notice that it is not necessary to sum in the above over all the $\binom{N}{r}$ r-tuples of k-tuples, but only over their index-sharing types. To illustrate this we take $k = 3$; for $r = 1$ there is no index-sharing possible and so, since all the sets $B_{i_1 i_2 i_3}$ are congruent, we have

$$K_1 = N \int \cdots \int_{B_{123}} \prod_{j=1}^{n} f(x_j) \, dx_1 \cdots dx_n \, .$$

This can be simplified on account of the special form of the integrand above; since the integral of $f(x)$ over A is 1 (f being a probability density) we have

$$K_1 = N \int_{H_{123}} f(x_1)f(x_2)f(x_3) \, dx_1 \, dx_2 \, dx_3 \, .$$

With $r = 2$ we have three types of index-sharing for two triples since there may be 0, 1, or 2 indices shared. In complete analogy to the incidence graphs of Figure 2 we have now incidence 2-graphs, shown together with the corresponding pairs of triples in Figure 3.

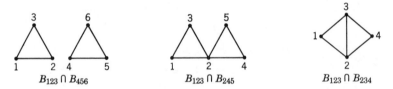

$B_{123} \cap B_{456}$ $B_{123} \cap B_{245}$ $B_{123} \cap B_{234}$

FIGURE 3.   Incidence hypergraphs and index-sharing for pairs of triples.

We return now to the case of general k and general r. Let $t(r, k)$ be the number of distinct incidence hypergraphs (k-graphs) and let G_{ri} be the ith one, $i = 1, \ldots, t(r, k)$. As before, $v(G_{ri})$ is the number of vertices of G_{ri}, and $N_{rik}(n)$ is the number of distinct k-graphs of the type of G_{ri} on the n vertices. The cluster integral $I(G_{ri})$ is defined for the incidence k-graph G_{ri} in the same way as $I(G)$ was defined in (11) with respect to an incidence graph G. We have now a different integrand instead of 1 in (11), namely the product $\prod_{j=1}^{n} f(x_j)$, and we generalize the conditions (a) and (b) following (11). Finally, the cluster integral $I(G_{ri})$ is factored into a product of cluster integrals over the connected components of G_{ri}. Putting together (15), (16), and (17) we have

$$P(A, a, f, n, k) = 1 - \sum_{r=1}^{\binom{n}{k}} \sum_{i=1}^{t(r, k)} (-1)^{r+1} N_{rik}(n) \prod_{C \subseteq G_{ri}} I(C) \qquad (18)$$

where C runs over all the connected components of the incidence k-graph G_{ri}.

While (18) provides a solution to our problems, it is clear that evaluating P is a matter of considerable difficulty. First, we have the combinatorial questions of determining the types of incidence hypergraphs G_{ri} and the numbers $t(r, k)$ and $N_{rik}(n)$. Then arises the analytical-numerical problem of evaluating the cluster integrals. Indeed, it might be questioned whether (18) constitutes the solution to our problem in any but a formal sense. However, here we must mention a very important lucky circumstance. It follows from the inclusion-exclusion formula (6a) that if the measure $m(X)$ is nonnegative then the inclusion-exclusion series is alternating. That is, in any such series of the type (7) we have

$$K_1 \geq K_2 \geq \cdots \geq K_N \geq 0.$$

Also, practice shows that in many problems of our type one wants the probability $P(A, a, f, n, k)$ for the case when a is very much smaller than the size of A. It turns out then that for the values of n and k which are not too large, the first few terms of (18) give a good approximation to the true value of P. What is equally important, the alternating series behavior of the inclusion-exclusion makes it possible to estimate the error of approximation.

BIBLIOGRAPHY

[1] André, D., *Comptes Rendus*, **88**, 965–967 (1879).

[2] Archimedes, *Works* (ed. T. L. Heath), Dover, New York, 1959.

[3] Artin, E., *The Gamma Function*, Holt, Rinehart and Winston, New York, 1964.

[4] Bell, E. T., *Ann. of Math.*, **35**, 258–277 (1934).

[5] Bender, E. and Goldman, J., *Combinatorial Analysis*, Holt, Rinehart and Winston (to appear).

[6] Bonnesen, T. and Fenchel, W., *Theorie der Konvexen Körper*, Chelsea, New York, 1948.

[7] Boyd, D. W., *Math. of Comp.*, **24**, 697–704 (1970).

[8] Boyd, D. W., *Aequationes Math.*, **7**, 182–193 (1971).

[9] Brooks, R. L. Smith, C. A. B., Stone, A. H., and Tutte, W. T., *Duke J. of Math.*, **7**, 312–340 (1940).

[10] Bromwich, T. J. I., *An Introduction to the Theory of Infinite Series*, 2nd ed., Macmillan, New York, 1942.

[11] Bruijn, N. G. de, *Asymptotic Methods in Analysis*, North-Holland, Amsterdam, 1958.

[12] Chern, S. S., in *Studies in Mathematics* (ed. S. S. Chern), vol. 4, Math. Assoc. of America and Prentice-Hall, 1967.

[13] Cockayne, E., *SIAM J. of Appl. Math.*, **15**, 1511–1516 (1967).

[14] Coddington, E. A. and Levinson, N., *Theory of Ordinary Differential Equations*, McGraw-Hill, New York, 1955.

[15] Courant, R. and Hilbert, D., *Methoden der Mathematischen Physik*, vol. 1, Springer, New York, 1931.

[16] Coxeter, H. S. M., *Introduction to Geometry*, Wiley, New York, 1961.

[17] Davis, H. T., *Introduction to Nonlinear Differential and Integral Equations*, Dover, New York, 1962.

[18] Dehn, M., *Math. Annalen*, **55**, 465–478 (1902).

[19] Dienes, P., *The Taylor Series*, Dover, New York, 1957.

[20] Dörrie, H., *Hundred Great Problems of Elementary Mathematics*, 2nd ed., Dover, New York, 1965.

[21] Edwards, J., *A Treatise on the Integral Calculus*, vols. 1 and 2, Macmillan, New York, 1921.

[22] Ford, L. R., *Automorphic Functions*, Chelsea, New York, 1951.

[23] Geppert, H., *Math. Annalen*, **107**, 387–399 (1932).

[24] Geppert, H., *Math. Annalen*, **108**, 197–207 (1933).

[25] Gleason, A., *Bull. Amer. Math. Soc.*, **55**, 446–449 (1949).

[26] Goursat, E., *A Course in Mathematical Analysis*, vol. 1, Ginn, Lexington, Mass., 1904.

[27] Hardy, G. H., *Ramanujan*, Cambr. Univ. Press, New York, 1940.

[28] Hardy, G. H. and Wright, E. M., *Introduction to the Theory of Numbers*, 4th ed., Oxf. Univ. Press, New York, 1968.

[29] Hausdorff, F., *Math. Annalen*, **94**, 244–247 (1925).

[30] Hermite, C., *Comptes Rendus*, **48**, 508–509 (1858).

[31] Hille, E., and Phillips, R. S., *Functional Analysis and Semigroups*, Amer. Math. Soc. Colloq. **31**, 1957.

[32] Hölder, O., *Math. Annalen*, **28**, 1–13 (1887).

[33] Hopcroft, J. E. and Kerr, L. R., in *Proc. Tenth Symp. on Switching and Automata*, 1969.

[34] Ivanoff, V. F., *Amer. Math. Monthly*, **65**, 212 (1958).

[35] King, L. V., *On the Direct Numerical Calculation of Elliptic Functions and Integrals*, Cambr. Univ. Press, New York, 1924.

[36] Kleene, S. C., *Introduction to Metamathematics*, Van Nostrand, New York, 1952.

[37] Klein, F., *Lectures on the Icosahedron*, 2nd ed., Dover, New York, 1956.

[38] Knopp, K., *Theory and Application of Infinite Series*, Blackie and Son, Glasgow, 1928.

[39] Landau, E., *Vorlesungen uber Zahlentheorie*, vol. 2, Chelsea, New York, 1947.

[40] Lech, C., *Ark. Mat.*, **2**, 417–421 (1953).

[41] LeVeque, W. J., *Topics in Number Theory*, vol. 2, Addison-Wesley, Reading, Mass., 1955.

[42] Lewin, L., *Dilogarithms and Associated Functions*, Macdonald, London, 1958.

[43] Mahler, K., *Cambr. Phil. Soc. Proc.*, **52**, 39–48 (1956).

[44] Maxwell, J. C., *Electricity and Magnetism*, Dover, New York, 1948.

[45] Melzak, Z. A., *Canad. Math. Bull.*, **2**, 175–180 (1958).

[46] Melzak, Z. A., *Canad. J. of Math.*, **12**, 20–26 (1960).

[47] Melzak, Z. A., *Math. of Comp.*, **23**, 169–172 (1969).

[48] Melzak, Z. A., *Bell Syst. Tech. J.*, **47**, 1105–1129 (1968).

[49] Meschkowski, H., *Ungeloste und Unlosbare Probleme der Geometrie*, Friedr. Vieweg, Braunschweig, 1960.

[50] Milnor, J., *Morse Theory*, Princeton Univ. Press, Princeton, N.J., 1963.

[51] Montgomery, D., and Zippin, L., *Ann. of Math.*, **56**, 213–241 (1952).

[52] Montgomery, D. and Zippin, L., *Topological Transformation Groups*, Interscience, New York, 1955.

[53] Moon, J. W., in *Seminar on Graph Theory* (ed. F. Harary), Holt, Rinehart and Winston, New York, 1967.

[54] Moore, E. F., in *Sequential Machines* (ed. E. F. Moore), Addison-Wesley, Reading, Mass., 1964.

[55] Nielsen, N., *Handbuch der Theorie der Gammafunktion*, Chelsea, New York, 1965.

[56] Niven, I., *Diophantine Approximation*, Interscience, New York, 1963.

[57] Oppenheim, A., in *Studies in Pure Mathematics* (ed. L. Mirsky), Academic Press, New York, 1971.

[58] Ostrowski, A., *Comm. Math. Helv.*, **18**, 283–308 (1946).

[59] Prim, R. C., *Bell. Syst. Tech. J.*, **31**, 1398–1401 (1957).

[60] Prüfer, H., *Arch. Math. Phys.*, **27**, 742–744 (1918).

[61] Riordan, J., *An Introduction to Combinatorial Analysis*, Wiley, New York, 1958.

[62] Riordan, J., *Combinatorial Identities*, Wiley, New York, 1968.

[63] Ritt, J. F., *Trans. Amer. Math. Soc.*, **27**, 68–90 (1925).

[64] Ritt, J. F., *Integration in Finite Terms*, Columbia Univ. Press, New York, 1948.

[65] Santalo, L. A., *Introduction to Integral Geometry*, Hermann et Cie., Paris, 1953.

[66] Scheid, F., *Theory and Problems of Numerical Analysis*, Schaum McGraw-Hill, New York, 1968.

[67] Siegel, C. L., *Tohoku Math. J.*, **20**, 26–31 (1921).

[68] Siegel, C. L., *Transcendental Numbers*, Princeton Univ. Press, Princeton, N.J., 1949.

[69] Sierpinski, W., *General Topology*, 2nd ed., Univ. of Toronto Press, Toronto, 1956.

[70] Strassen, V., *Numer. Math.*, **13**, 354–356 (1969).

[71] Struik, D. J., *Differential Geometry*, Addison-Wesley, Reading, Mass., 1950.

[72] Titchmarsh, E. C., *Proc. London Math. Soc.*, **28**, 70–80 (1929).

[73] Titchmarsh, E. C., *The Theory of the Riemann Zeta Function*, Oxf. Univ. Press, New York, 1951.

[74] Uspensky, J. V. and Heaslet, M. A., *Elementary Number Theory*, McGraw-Hill, New York, 1939.

[75] Waerden, B. L. van der, *Modern Algebra*, vol. 1, Ungar, New York, 1949.

[76] Watson, G. N., *A Treatise on the Theory of Bessel Functions*, 2nd ed., Cambr. Univ. Press, New York, 1958.

[77] Whittaker, E. T. and Watson, G. N., *A Course of Modern Analysis*, Amer. ed., Macmillan, New York, 1948.

[78] Winograd, S., *Lin. Alg. and Appl.*, **4**, 381–388 (1971).

[79] Winograd, S., *Lin. Alg. and Appl.*, **4**, 377–379 (1971).

INDEX

265